THE MANY-BODY PROBLEM

THE MANY-BODY PROBLEM

Mallorca International School of Physics
August 1969

Director: L. M. Garrido
Professor of Theoretical Physics
University of Barcelona

Edited by

A. Cruz
University of Zaragoza

and

T. W. Preist
University of Exeter

Springer Science+Business Media, LLC 1969

ISBN 978-1-4899-6164-8 ISBN 978-1-4899-6319-2 (eBook)
DOI 10.1007/978-1-4899-6319-2

Library of Congress Catalog Card Number 77-94344

Originally published by Plenum Press. New York 1969.
Softcover reprint of the hardcover 1st edition 1969
United Kingdom edition published by Plenum Press, London
A Division of Plenum Publishing Company, Ltd.
Donington House, 30 Norfolk Street, London W.C. 2, England

PREFACE

The Mallorca International School of Physics was initiated with the aim of fostering new fields of research in physics in the Spanish Universities as well as consolidating those fields in which research is already being pursued. In addition, a School of this character would benefit many students from other countries.

Students of many countries had the opportunity of listening to nearly 60 especially prepared lectures, by physicists, specialists in their field, as well as making personal contact with them (and amongst themselves) to discuss scientific problems of common interest. As a result, it is reasonable to expect that a regularly constituted school, complementary to others already established in Europe, would benefit the staff and students of the Spanish Universities within a period of two or three years.

The subject of The Many-Body Problem was most appropriate for the first school since it is a topic of immediate interest in many branches of Physics - Statistical Mechanics, Nuclear Physics, Solid State Physics and Plasma Physics. The School was fortunate in obtaining an eminent collection of lecturers and I am most grateful that they could spare some of their valuable time to give the new School such a good start. All the lectures appear in the Proceedings with the exception of those by Professor L. Van Hove and Professor P.C. Martin, whose lectures have previously been published. Consequently, the School and these Proceedings will be of benefit to many.

The topicality and attractiveness of the programme was reflected in the large number of applicants to attend the School and some were no doubt additionally influenced by the siting of the School - on a beautiful island, with an international airport, at a world famous holiday resort renowned for sporting and relaxation facilities.

As I have said, it is essential that the School should be a regular one if it is to produce lasting benefit to Spanish scientists and I hope that the Spanish authorities will continue to look on this endeavour in a sympathetic way.

Finally, I have much pleasure in giving my sincerest thanks to many people who helped to transform the School from an idea to a reality:

Professor J.L. Villar-Palasí, Minister of Education of Spain for sponsoring the School.

Dr. R. Diez-Hochleitner, Secretary General of the Ministry of Education for receiving the project of the School with such enthusiasm.

Professor F. Rodriguez, Director General of Universities in Spain and Dr. A. de Juan-Abad for providing the necessary finance.

Professor G. Tomás , Rector of the Estudio General Luliano, where the lectures were held.

The Majorcan Authorities, who contributed to making the School a success.

Professor Sir Rudolph Peierls, F.R.S. (University of Oxford) for continuous moral support in the organization of the School.

Professor N. Kemmer, F.R.S. (University of Edinburgh), who provided so much useful information on the running of a School of this nature.

Dr. A. Cruz (University of Zaragoza) and Dr. T.W. Preist (University of Exeter) co-editors of the Proceedings.

And finally, Mrs. R.W. Chester (University of Edinburgh) and Miss H.H. Ostermann (University of Barcelona) for their heroic efforts in the typing of the manuscript and the organization of the School.

L.M. GARRIDO
Professor of Theoretical Physics
University of Barcelona

EDITORS' NOTE

The lecture notes were prepared by the lecturers themselves before or during the School. The final manuscript was typed during and after the School had finished. As a result most of the lecturers were unable to see it in its final form and consequently the responsibility for any errors and misprints lies with the editors.

PARTICIPANTS OF THE 1969 SUMMER SCHOOL

LECTURERS

Prof. L.J. Boya, Valladolid
Prof. E.R. Caianiello, Naples
Dr. C.B. Dover, Heidelberg
Prof. C.P. Enz, Geneva
Prof. I. Fujiwara, Trondheim
Prof. L. van Hove, CERN
Prof. N.J. Horing, Stevens I.T.
Prof. P.C. Martin, Harvard
Prof. W. Thirring, CERN
Prof. E.J. Verboven, Nijmegen

PARTICIPANTS

Mr. J. Arponen, Nordita
Mr. L.C. Balbas, Valladolid
Dr. R. von Baltz, Karlsruhe
Dr. A. Baracca, Florence
Dr. A. Bernalte, Barcelona
Dr. J. Biel, Zaragoza
Prof. H. Bolton, Oxford
Dr. J.V. Bonet, Buffalo
Dr. E. Borchi, Florence
Mr. J. Bordas, Cambridge
Dr. A. Bramón, Barcelona
Mr. S. Cendejas, Mexico
Mr. A.A. Cottey, Norwich
Mr. M. Droz, Geneva
Mr. H. Duarte, Coimbra
Dr. H. Falk, New Jersey

Mr. J. Fernandez, Valladolid
Mr. J.J. Forney, Geneva
Mr. J. Friedman, New York
Dr. F. Gascón, Sevilla
Prof. F. Gautier, Strasbourg
Mr. J. Gomis, Barcelona
Mr. J.A. Grifols, Barcelona
Dr. F. Hahne, Heidelberg
Dr. H. Hermeking, Cologne
Dr. G. Iadonisi, Naples
Mr. R.C. Jones, Manchester
Dr. D. Kobe, Texas
Dr. J. Krieger, New York
Mr. H. Kunz, Geneva
Dr. M. Lombardero, Madrid
Mr. F. Lopez, Valladolid
Mr. Ch. Lydén, Gothenburg
Mr. M. Mac Innes, Syracuse
Dr. M. de Maria, Rome
Dr. M. Marinaro, Naples
Mr. J. Marro, Barcelona
Mr. M.A. Martinez, Mexico
Mr. F. Mauricio, San Sebastián
Dr. P. Meier, Zürich
Mr. A. Montoto, Barcelona
Mr. J.M. Navarro, Madrid
Mr. E. Pajanne, Gothenburg
Dr. F. Palumbo, Rome
Mr. M. Paramio, Madrid
Mr. J.M. Prado, Cambridge
Mr. P. Puigdomenech, Barcelona
Mr. H. Rayo, Barcelona
Dr. H. Rietschel, Karlsruhe

Mr. C. Sala, Barcelona
Mr. V. Samathiyakanit, Gothenburg
Mr. R. Sandstrom, Stockholm
Mr. A. Santistebán, Madrid
Prof. E. Santos, Valladolid
Dr. G. Sartori, Padua
Mr. A. Sendino, Valladolid
Dr. T. Siskens, Leiden
Dr. J. Speth, Munich
Dr. B. Tirozzi, Rome
Dr. J.W. Tucker, Sheffield
Dr. J.N. Urbano, Coimbra
Mr. V. van Doren, Georgia
Dr. R. Watts-Tobin, Lancaster
Dr. A. Wehrl, CERN
Mr. F. Weinstein, Princeton
Dr. P. Wölfle, Munich
Mr. F. Yndurain, Madrid
Mr. J. Zofka, Prague

DIRECTOR

Prof. L.M. Garrido, Barcelona

EDITORS

Dr. A. Cruz, Zaragoza
Dr. T.W. Preist, Exeter

SECRETARIAT

Mrs. R.W. Chester, Edinburgh
Miss H.H. Ostermann, Barcelona

CONTENTS

GREEN FUNCTIONS APPLIED TO PHONON PROBLEMS

Charles P. Enz
Institut de Physique Théorique
Ecole de Physique, CH-1211 GENEVE 4

I. INTRODUCTION TO DYNAMICS AND TO GREEN FUNCTIONS

1. CRYSTAL AND LIQUID HAMILTONIANS

1.1 Crystal Hamiltonian in Adiabatic Approximation[1,2,3]

We introduce the following notation: N is the number of unit cells in the crystal volume V as defined by periodic boundary conditions, and $\vec{R}_I$ $(I = 1,\dots N)$ are the cell positions. Each cell contains B basis ions ($B = 1$ for a Bravais lattice) with mass M_b and equilibrium position $\vec{R}^o_{Ib}$ $(b = 1,\dots B)$; we fix $\vec{R}^o_{I1} = \vec{R}_I$. The instantaneous positions of the ions are given by $R = \left\{\vec{R}_{Ib} = \vec{R}^o_{Ib} + \vec{u}_{Ib}\right\}$ where $\vec{u}_{Ib}$ are their displacements. Z is the number of valence electrons per cell at temperature $T = 0$ ($Z = 0$ for an insulator) and $r = \left\{r_e\right\}$ $(e = 1,\dots, ZN)$ are the positions of the valence electrons (mass m).

Since m/M_b is always very small $(< 10^{-3})$ the electrons move so fast as to screen the long range Coulomb interaction of the ions. Then the Hamiltonian may be written

$$H = H_{ion} + H_{el} \tag{1.1.1}$$

$$H_{ion} = \sum_{I,b} \frac{\vec{P}_{Ib}^{\,2}}{2M_b} + U(R) \tag{1.1.2}$$

$$H_{el}(R) = \sum_e \left(\frac{\vec{p}_e^{\,2}}{2m} + W(\vec{r}_e, R)\right) + H_{coul} \tag{1.1.3}$$

where H_{coul} is the Coulomb interaction between electrons. In the adiabatic or Born-Oppenheimer approximation the electron dynamics is solved for fixed instantaneous ion positions

$$H_{el}(R)|\ell,R\rangle = E_\ell(R)|\ell,R\rangle \tag{1.1.4}$$

$$\langle\ell',R|\ell,R\rangle = \delta_{\ell\ell'} \tag{1.1.4$'$}$$

and the effect of the ionic kinetic energy on these states is neglected,

$$\langle\ell',R|H|\ell,R\rangle \cong (H_{ion} + E\,(R))\,\delta_{\ell\ell'} \tag{1.1.5}$$

If the displacements $\vec{u}_{Ib}$ are small compared to the interatomic distances, (1.1.4) may first be solved to zeroth order in the $\vec{u}_{Ib}$ by a Hartree-Fock procedure

$$H_{el}(R^o) = H_{HF} + H_{el\text{-}el} - \tfrac{1}{2}\sum_e V_{HF}(\vec{r}_e) \tag{1.1.6}$$

where $V_{HF}(\vec{r}_e)$ is the self-consistent field determined such that the residual electron-electron interaction

$$H_{el\text{-}el} = H_{coul} - \tfrac{1}{2}\sum_e V_{HF}(r_e) \tag{1.1.7}$$

does not contribute to the energy of the Hartree-Fock ground state $|\phi_o\rangle$, $\langle\phi_o|H_{el\text{-}el}|\phi_o\rangle = 0$.

$$H_{HF} = \sum_e \left(\frac{\vec{p}_e^{\,2}}{2m} + V(\vec{r}_e)\right) \tag{1.1.8}$$

where

$$V(\vec{r}) = W(\vec{r},R) + V_{HF}(\vec{r}) \tag{1.1.9}$$

is the periodic potential, defines free quasi-particles

Their eigenstates and eigenvalues are Bloch functions $|k\rangle$ and energy bands ε_k^o with $k = (\vec{k}, s, n)$. Here $\vec{k}$ is a vector in the reduced zone, s the spin index and n the band index. In second quantization

$$H_{HF} = \sum_k \varepsilon_k^o a_k^+ a_k \tag{1.1.10}$$

$$a_k|\phi_o\rangle = 0, \quad a_k^+|\phi_o\rangle = |k\rangle; \ \varepsilon_k^o > 0 \tag{1.1.11}$$

describes an electron and

$$a_k^+|\phi_o\rangle = 0, \quad a_k|\phi_o\rangle = |k\rangle; \ \varepsilon_k^o > 0 \tag{1.1.11'}$$

describes a hole. The Fermi energy is given by $\varepsilon_{k_F}^o = 0$ where k_F is the Fermi momentum (we put $\hbar = 1$ throughout).

According to (1.1.5) the ionic system may be described by an effective Hamiltonian

$$\tilde{H}_{ion} = H_{ion} + E_\ell(R) = \sum_{I,b} \frac{\vec{P}_{Ib}^{\,2}}{2M_b} + \tilde{U}(R) \tag{1.1.12}$$

which still depends on the electronic state ℓ. However, in the adiabatic approximation the ionic motion does not induce transitions of the state ℓ so that is fixed.

A Taylor series expansion of $\tilde{U}(R)$, using the notation $x = (I,b,i)$, where $i = 1,2,3$ labels the cartesian vector components, and

$$U_{x_1 \ldots x_n} \equiv \left(\frac{\partial^n \tilde{U}}{\partial R_{x_1} \ldots \partial R_{x_n}}\right)_{R_o} \tag{1.1.13}$$

gives

$$\tilde{U}(R) = \tilde{U}(R^o) + \sum_x U_x u_x + \frac{1}{2!} \sum_{x_1 x_2} U_{x_1 x_2} u_{x_1} u_{x_2} + \ldots \tag{1.1.14}$$

With the equilibrium condition at $T = 0$

$$U_x = 0; \quad \text{all } x \tag{1.1.15}$$

we have

$$\widetilde{H}_{ion} = \widetilde{U}(R^o) + H_{harm} + H_{anh} \tag{1.1.16}$$

where $(M_x \equiv M_b)$

$$H_{harm} = \sum_x \frac{P_x^2}{2M_x} + \frac{1}{2} \sum_{x_1 x_2} U_{x_1 x_2} u_{x_1} u_{x_2} \tag{1.1.17}$$

describes undisturbed harmonic oscillations and

$$H_{anh} = \sum_{n \geqslant 3} H_n \tag{1.1.18}$$

with

$$H_n = \frac{1}{n!} \sum_{x_1 \ldots x_n} U_{x_1 \ldots x_n} u_{x_1} \ldots u_{x_n} \tag{1.1.19}$$

the anharmonic interactions.

It should be noted that for the lightest atoms (m/M_b largest) crystal dynamics is not appropriately described by an immediate application of such a development. In this case the undisturbed oscillations have first to be determined selfconsistently by a Hartree method. This is the problem of the <u>quantum crystals</u> which will not be treated here.

<u>Normal coordinates</u> are introduced by[4)]

$$\left.\begin{aligned} u_x &= \sum_q \Psi_q(x)(M_x \omega_q^o)^{-\frac{1}{2}} Q_q \\ P_x &= \sum_q \Psi_q(x)(M_x \omega_q^o)^{\frac{1}{2}} P_{-q} \end{aligned}\right\} \tag{1.1.20}$$

Here

$$\Psi_q(x) \equiv N^{-\frac{1}{2}} e_{bi}(q) e^{i\vec{q}.\vec{R}_I} = \Psi^*_{-q}(x) \tag{1.1.20'}$$

and $\pm q = (\pm \vec{q}, \mu)$ where $\vec{q}$ is a vector in the reduced zone and $\mu = 1, \ldots, 3B$, labels the modes and polarizations of the phonons. The frequencies $\omega_q^o = \omega_{-q}^o$ and polarization vectors

$\vec{e}_b(q) = \vec{e}_b^{\,*}(-q)$ are determined by the eigenvalue equation

$$\sum_{b'i'} (D_{bi,b'i'}(\vec{q}) - \omega^{o}_{q}\, \delta_{bi,b'i'}) e_{b'i'}(q) = 0 \tag{1.1.21}$$

where

$$D_{bi,b'i'}(\vec{q}) = \sum_{I} (M_b M_{b'})^{-\frac{1}{2}}\, U_{xx'}\, e^{-i\vec{q}.(\vec{R}_I - \vec{R}_{I'})} \tag{1.1.21'}$$

and the normalization condition

$$\frac{1}{B} \sum_{b} \vec{e}^{\,*}_{b}(\vec{q},\mu).\vec{e}_{b}(\vec{q},\mu') = \delta_{\mu\mu'}. \tag{1.1.22}$$

Introducing creation and annihilation operators by

$$\left.\begin{aligned} Q_q &= Q^{+}_{-q} = \frac{1}{\sqrt{2}} (b_q + b^{+}_{-q}) \\ P_{-q} &= P^{+}_{q} = \frac{1}{i\sqrt{2}} (b_q - b^{+}_{-q}) \end{aligned}\right\} \tag{1.1.23}$$

we have

$$H_{harm} = \tfrac{1}{2} \sum_{q} \omega^{o}_{q} (P_q P^{+}_{q} + Q_q Q^{+}_{q}) = \sum_{q} \omega^{o}_{q} (b^{+}_{q} b_q + \tfrac{1}{2}) \tag{1.1.24}$$

The interaction terms become

$$H_n = \frac{1}{n!} \sum_{q_1 \ldots q_n} C^{(n)}_{q_1 \ldots q_n} Q_{q_1} \ldots Q_{q_n} \tag{1.1.25}$$

where

$$C^{(n)}_{q_1 \ldots q_n} = C^{(n)\,*}_{-q_1 \ldots -q_n}$$

$$\sum_{x_1 \ldots x_n} U_{x_1 \ldots x_n} \prod_{j=1}^{n} \left[(\omega^{o}_{q_j} M_{b_j})^{-\frac{1}{2}} \psi_{q_j}(x_j) \right] \tag{1.1.26}$$

These coupling functions $C^{(n)}_{(q)}$ contain a crystal momentum conservation factor $\Delta(\vec{q}_1 + \ldots + \vec{q}_n)$ where

$$\Delta(\vec{p}) \equiv \frac{1}{N} \sum_{I} e^{i\vec{p}.\vec{R}_I} = \sum_{\vec{K}} \delta_{\vec{p},\vec{K}} \tag{1.1.27}$$

and $\vec{K}$ is a vector of the reciprocal lattice. $\vec{K} = 0$ gives normal processes, $\vec{K} \neq 0$ gives Umklapp processes.

The electron-phonon interaction is obtained by a development of $H_{el}(R)$ in the $\vec{u}_{Ib}$

$$H_{el}(R) = H_{el}(R^{o}) + H_{el-ph} . \qquad (1.1.28)$$

To first order

$$H_{el-ph} = \sum_{Ib} \left(\frac{\partial W}{\partial \vec{R}_{Ib}}\right)_{R_o} \cdot \vec{u}_{Ib} \qquad (1.1.29)$$

or in second quantization of the electrons

$$H_{el-ph} = \sum_{kk'q} g_{k'kq}\, a^{+}_{k'}\, a_k\, Q_q \qquad (1.1.30)$$

where

$$g_{k'kq} = g^{*}_{kk'-q}$$

$$= \sum_{Ib} (NM_b \omega^{o}_{q})^{-\frac{1}{2}} \langle k' | \frac{\partial W}{\partial \vec{R}_{Ib}} | k \rangle \vec{e}_b(q) e^{i\vec{q}.\vec{R}_I} \qquad (1.1.31)$$

Again $g_{k'kq}$ contains a crystal momentum conservation factor $\Delta(\vec{k} + \vec{q} - \vec{k}')$.

In the approximation of two-body forces and rigid ions

$$W(\vec{r}, R) = \sum_{Ib} \varphi_b(\vec{r} - \vec{R}_{Ib})$$

and taking plane waves for the states $|k\rangle$ equation (1.1.31) becomes

$$g_{k'kq} = \sum_{\vec{K}} \gamma_q(\vec{K})\, \delta_{\vec{k}+\vec{q}-\vec{k}', \vec{K}}\, \delta_{s's} \qquad (1.1.32)$$

with

$$\gamma_q(\vec{K}) = \gamma^{*}_{-q}(-\vec{K})$$

$$= -i\frac{N}{V} \sum_{b} (NM_b \omega^{o}_{q})^{-\frac{1}{2}} (\vec{q}+\vec{K}).\vec{e}_b(q) \int_V d^3r\, \varphi_b(\vec{r}) e^{-i(\vec{q}+\vec{K}).\vec{r}} \qquad (1.1.33)$$

This shows that normal processes ($\vec{K} = 0$) only couple longitudinal phonons ($\vec{e}_b(q)$ parallel to $\vec{q}$).

The hierarchy within the adiabatic approximation of the various hamiltonians is given by the power of the small quantity

$$\kappa \equiv \left(\frac{m}{M}\right)^{\frac{1}{4}} \lesssim 0.1 \tag{1.1.34}$$

where M is an average ion mass. One finds[5)]

$$H_{HF} \propto \kappa^{0}, \quad H_{el-ph} \propto \kappa^{1}, \quad H_{harm} \propto \kappa^{2}, \quad H_{n} \propto \kappa^{n}. \tag{1.1.35}$$

1.2 The "Hamiltonian" of Quantum Hydrodynamics[6)]

Following Landau the internal energy at very low temperature of a quantum fluid of bosons can be cast into a form analogous to the phonon Hamiltonian (1.24, 25) by choosing as canonically conjugate variables the density variation ρ' and the velocity potential ϕ. This, of course, is only a model since starting from a many-body description of the atoms of the fluid ρ' and ϕ turn out not to be canonically conjugate.

The internal density is

$$h(\vec{r}) = \tfrac{1}{2}\rho\vec{W}^2 - \frac{1}{v(r)}\int_{v_0}^{v(r)} (p-p_0)\,dv_1(r) \tag{1.1.36}$$

where p is the pressure, $v(r)$ the volume per particle at $\vec{r}$ and v_0 its mean value. With $\vec{W} = -\nabla\phi$ and the mass density $\rho(\vec{r}) = M/v(\vec{r}) = \rho_0 + \rho'$, $\rho_0 \equiv M/v_0$,

$$\begin{aligned} h(\vec{r}) &= \tfrac{1}{2}\rho(\nabla\phi)^2 + \rho\int_{\rho_0}^{\rho}\left[p(\rho_1) - p(\rho_0)\right]\frac{d\rho_1}{\rho_1^2} \\ &= h_0 + h_3^k + \sum_{n\geq 3} h_n \end{aligned} \tag{1.1.37}$$

where

$$h_0 = \frac{\rho_0}{2}(\nabla\phi)^2 + \frac{c^2}{2\rho_0}(\rho')^2 \tag{1.1.38}$$

is the harmonic part,

$$h_3^k = \tfrac{1}{2}\rho'(\nabla\phi)^2 \tag{1.1.39}$$

the anharmonic term of the kinetic energy density and

$$h_3 = \frac{2u-1}{6}\cdot\frac{c^2}{\rho_0^2}(\rho')^3 \tag{1.1.40}$$

$$h_4 = (1 - 3u + 2u^2 + \rho_0\frac{\partial u}{\partial\rho_0})\,\frac{c^2}{12\rho_o}(\rho')^4 \tag{1.1.41}$$

etc. the anharmonic parts of the potential energy density. Here

$$c = \left[(\frac{\partial p}{\partial\rho})_{T,\rho_0}\right]^{\frac{1}{2}} \tag{1.1.42}$$

is the hydrodynamic sound velocity and

$$u = \frac{\partial\log c}{\partial\log\rho_0} \tag{1.1.43}$$

the Grüneisen constant.

Introducing normal coordinates by

$$\rho' = \sum_{\vec{q}}(\frac{q\rho_0}{Vc})^{\frac{1}{2}}\,e^{i\vec{q}.\vec{r}}\,Q_{\vec{q}} \tag{1.1.44}$$

$$= -\sum_{\vec{q}}(\frac{c}{qV\rho_0})^{\frac{1}{2}}\,e^{i\vec{q}.\vec{r}}\,P_{-\vec{q}} \tag{1.1.45}$$

(where $V = Nv_0$ is the total volume), and relating $Q_{\vec{q}}$ and $P_{\vec{q}}$ to emission and absorption operators as in (1.1.23) the "Hamiltonian"

$$H = \int_V d^3r h(r) = H_0 + H_3^k + \sum_{n\geq 3} H_n \tag{1.1.46}$$

takes the form analogous to (1.1.24, 25).

$$H_0 = \frac{1}{2} \sum_{\vec{q}} \omega_q (P_{\vec{q}} P^+_{\vec{q}} + Q_{\vec{q}} Q^+_{\vec{q}}) \qquad (1.1.47)$$

$$H_3^k = \frac{1}{3!} \sum_{\vec{q}_1\vec{q}_2\vec{q}_3} C^k_{\vec{q}_1\vec{q}_2\vec{q}_3}$$

$$\cdot \frac{1}{2}(Q_{\vec{q}_1} P^+_{\vec{q}_2} P^+_{\vec{q}_3} + P^+_{\vec{q}_2} P^+_{\vec{q}_3} Q_{\vec{q}_1}) \qquad (1.1.48)$$

$$H_n = \frac{1}{n!} \sum_{\vec{q}_1 \ldots \vec{q}_n} C^{(n)}_{\vec{q}_1 \ldots \vec{q}_n} Q_{\vec{q}_1} \ldots Q_{\vec{q}_n} \qquad (1.1.49)$$

where

$$\omega_q = cq \; ; \quad q \equiv |\vec{q}| \qquad (1.1.50)$$

$$C^{(n)}_{\vec{q}_1 \ldots \vec{q}_n} = N^{1-\frac{n}{2}} C_n \, (\omega_{q_1} \ldots \omega_{q_n})^{\frac{1}{2}} \delta_{\vec{q}_1 + \ldots + \vec{q}_n, 0} \qquad (1.1.51)$$

$$C_3 = \frac{2u - 1}{c\sqrt{M}}$$

$$C_4 = \frac{2}{c^2 M} \left(1 - 3u + 2u^2 + \rho_0 \frac{\partial u}{\partial \rho_o}\right) \qquad (1.1.52)$$

$$C^k_{\vec{q}_1\vec{q}_2\vec{q}_3} = - \frac{3}{2u-1} C^{(3)}_{\vec{q}_1\vec{q}_2\vec{q}_3} (\hat{q}_2 \cdot \hat{q}_3) \qquad (1.1.53)$$

and $\hat{q} = \vec{q}/q$.

2. REAL TIME GREEN FUNCTIONS[3,7]

2.1 Definitions

We define the time ordered Green function of two one-particle operators A and B by

$$G_{AB}(t-t') = -i \langle T(A(t)B(t')) \rangle \qquad (1.2.1)$$

or more generally

$$G_{A_1 \ldots A_m}(t_1 - t_m, \ldots, t_{m-1} - t_m) =$$

$$(-i)^{m/2} \langle T(A_1(t_1) \ldots A_m(t_m)) \rangle \qquad (1.2.1')$$

where

$$A(t) = e^{i(H-\mu N)t} A e^{-i(H-\mu N)t} \qquad (1.2.2)$$

is the Heisenberg representation of statistical mechanics. For time-dependent Hamiltonians (2.2) has to be replaced by

$$\frac{\partial A(t)}{\partial t} = i\left[H(t)-\mu N(t), A(t)\right]_- \qquad (1.2.2')$$

$N = \sum_k a_k^+ a_k$ is the number operator of the conserved particles, (i.e. the electrons but not the phonons), which commutes with all the terms of the hamiltonian H as described in section 1. μ is the associated chemical potential.

T is the real-time ordering operator

$$T(A_1(t_1) \ldots A_m(t_m)) = \sum_{\text{permutations}} (\pm 1)^{\varepsilon_p} \; .$$

$$\cdot A_{i_1}(t_{i_1}) \Theta(t_{i_1} - t_{i_2}) A_{i_2}(t_{i_2}) \ldots \Theta(t_{i_{m-1}} - t_{i_m}) A_{i_m}(t_{i_m}) \qquad (1.2.3)$$

where $\Theta(t)$ is the step function (+1 for $t > 0$, 0 for $t < 0$) and the upper/lower sign applies to boson-fermion operators.

The average in (1.2.1) is taken in the grand canonical ensemble

$$\rho = e^{\beta(\Omega - H + \mu N)} \qquad (1.2.4)$$

where $\beta = 1/T$ (we put $k_B = 1$ throughout)

$$\langle A \rangle = \mathrm{Tr}(\rho A) , \qquad (1.2.5)$$

the thermodynamic potential Ω being determined by the condition

$$\mathrm{Tr}\,\rho = 1 . \qquad (1.2.5')$$

The one-electron and one-phonon Green functions in momentum space will be denoted by

$$G^{()}_{a_k\, a^+_{k'}}(t) \equiv G^{()}(k, t)\, \delta_{\vec{k}\,\vec{k}'} \qquad (1.2.6)$$

and

$$G^{()}_{Q_q\, Q^+_{q'}}(t) \equiv D^{()}(q, t)\, \delta_{\vec{q}\,\vec{q}'} \qquad (1.2.7)$$

In absence of magnetic interactions $G^{()}(k,t)$ is diagonal in the spin index whereas $D^{()}(q,t)$ is in general a matrix in the mode and polarization index μ.

In turn we can also take A and B in (1.2.1) in configuration space, i.e. to be the field operator $\psi(\vec{r}) = \sum_k a_k |k\rangle$ of the electrons or $\vec{u}_b(\vec{r})$, of the phonons (1.1.20).

Other useful functions are the <u>correlation functions</u> of Kadanoff and Baym[8])

$$G^{>}_{AB}(t-t') = -i\langle A(t)B(t')\rangle \qquad (1.2.8)$$

$$G^{<}_{AB}(t-t') = \mp i\langle B(t')A(t)\rangle \qquad (1.2.9)$$

and the <u>retarded Green function</u>

$$G^{ret}_{AB}(t-t') = -i\langle \mathcal{R}(A(t)B(t'))\rangle \qquad (1.2.10)$$

where $\mathcal{R}$ is the retardation operator defined for bosons/fermions by the retarded commutator/anticommutator

$$\mathcal{R}(A(t)B(t')) = \Theta(t-t')\left[A(t), B(t')\right]_{\mp} . \qquad (1.2.11)$$

These functions are related by

$$G^{ret}_{AB}(t) = G_{AB}(t) - G^{<}_{AB}(t) = \Theta(t)(G^{>}_{AB}(t) - G^{<}_{AB}(t)) \qquad (1.2.12)$$

and their Fourier transforms are defined by

$$G_{AB}^{()}(\omega) = \int dt\, G_{AB}^{()}(t)e^{i\omega t} . \qquad (1.2.13)$$

Writing the hamiltonian as

$$H = H_o + H_{int} \qquad (1.2.14)$$

where according to section 1.1 the unperturbed part H_o consists of H_{HF} and H_{harm}, (1.1.10, 24), and H_{int} contains the electron-electron, electron-phonon and phonon-phonon interactions, (1.1.7, 30, 25), we define unperturbed Green functions or propagator by

$$G_{AB}^{o}(t-t') = -i\langle T(A[t]B[t'])\rangle_o \qquad (1.2.15)$$

etc. Here the straight brackets denote the interaction representation

$$A[t] = e^{i(H_o-\mu N)t} A e^{-i(H_o-\mu N)t} \qquad (1.2.16)$$

For time-dependent Hamiltonians (1.2.16) has to be replaced by

$$\frac{\partial A[t]}{\partial t} = i\left[H_o[t] - N[t], A[t]\right]_- \qquad (1.2.16')$$

The average $\langle\ \rangle_o$ is taken in the unperturbed ensemble

$$\rho^o = e^{i(\Omega_o - H_o + \mu N)} \qquad (1.2.17)$$

where $\langle\rangle_o$ is determined by

$$\mathrm{Tr}\,\rho^o = 1 . \qquad (1.2.17')$$

One then finds for the unperturbed Fourier transformed time-ordered functions (1.2.6) and (1.2.7)

$$G^o(k,\omega) = \frac{1-f_k^o}{\omega - \varepsilon_k^o + i\delta} + \frac{f_k^{\,o}}{\omega - \varepsilon_k^o - i\delta} \qquad (1.2.18)$$

and

$$D^o(q,\omega) = \frac{(1+n_q^o)\omega_q^o}{\omega^2-(\omega_q^o - i\delta)^2} - \frac{n_q^o\,\omega_q^o}{\omega^2-(\omega_q^o + i\delta)^2} \qquad (1.2.19)$$

where δ is a positive infinitesimal;

$$f_k^0 = \langle a_k^+ a_k \rangle_0 = \frac{1}{e^{\beta \varepsilon_k^0} + 1} \equiv f^0(\varepsilon_k^0) \quad (1.2.20)$$

and

$$n_q^0 = \langle b_q^+ b_q \rangle_0 = \frac{1}{e^{\beta \omega_q^0} - 1} \equiv n^0(\omega_q^0) \quad (1.2.21)$$

are the unperturbed electron and phonon distribution functions.

Introducing the electron self-energy Σ and the phonon self-energy Π by the Dyson equations

$$G(k,\omega) = G^0(k,\omega) + G^0(k,\omega)\Sigma(k,\omega)G(k,\omega) \quad (1.2.22)$$

and

$$D(q,\omega) = D^0(q,\omega) + D^0(q,\omega)\Pi(q,\omega)D(q,\omega) \quad (1.2.23)$$

and making use of (1.2.18) and (1.2.19) we find

$$G(k,\omega) = \frac{1}{\omega - \varepsilon_k^0 - \Sigma(k,\omega)} \quad (1.2.24)$$

and

$$D(q,\omega) = \frac{\omega_q^0}{\omega^2 - \omega_q^{0\,2} - \omega_q^0 \Pi(q,\omega)} \quad (1.2.25)$$

2.2 Spectral Representation and Sum Rules

In the representation where H and N are diagonal

$$H|\Psi_s\rangle = E_s|\Psi_s\rangle$$
$$N|\Psi_s\rangle = N_s|\Psi_s\rangle \quad (1.2.26)$$
$$\rho|\Psi_s\rangle = \rho_s|\Psi_s\rangle$$

we find immediately

$$-\frac{1}{2\pi i} G^{>}_{AB}(\omega) \equiv C^{+}_{AB}(\omega) =$$
$$= \sum_{ss'} \rho_s \langle \Psi_s | A | \Psi_{s'} \rangle \langle \Psi_{s'} | B | \Psi_s \rangle \delta(\omega + E_{ss'} - \mu N_{ss'}) \tag{1.2.27}$$

and

$$+\frac{1}{2\pi i} G^{<}_{AB}(\omega) \equiv C^{-}_{AB}(\omega)$$
$$= \sum_{ss'} \rho_s \langle \Psi_s | B | \Psi_{s'} \rangle \langle \Psi_{s'} | A | \Psi_s \rangle \delta(\omega - E_{ss'} + \mu N_{ss'}) \tag{1.2.28}$$

($-/+$ for boson/fermion operator) where $E_{ss'} \equiv E_s - E_{s'}$ and $N_{ss'} \equiv N_s - N_{s'}$. On the other hand the Fourier transforms of the functions

$$\begin{aligned} G^{+}_{AB}(t) &= \theta(t)\, G^{>}_{AB}(t) \\ G^{-}_{AB}(t) &= \pm\, \theta(t)\, G^{<}_{AB}(t) \end{aligned} \tag{1.2.29}$$

($+/-$ for boson/fermion operators) are boundary values at $z = \omega + i\delta$ of the functions

$$G^{\pm}_{AB}(z) = \int_{-\infty}^{+\infty} d\omega' \, \frac{C^{\pm}_{AB}(\omega')}{z - \omega'} \tag{1.2.29'}$$

which are analytic in the z-plane cut along the real axis. (Continuation across the real z-axis can be achieved by writing $G^{\pm}_{AB}(z)$ as sum of two functions each of which has a cut on only half the real z-axis). It then follows from (1.2.12, 27 - 29') that

$$G_{AB}(\omega) = \int_{-\infty}^{+\infty} \frac{C^{+}_{AB}(\omega')d\omega'}{\omega - \omega' + i\delta} \pm \int_{-\infty}^{+\infty} \frac{C^{-}_{AB}(\omega')d\omega'}{\omega - \omega' - i\delta} \tag{1.2.30}$$

($+/-$ for boson/fermion operators) and

$$G^{ret}_{AB}(\omega) = \int_{-\infty}^{+\infty} \frac{C_{AB}(\omega')d\omega'}{\omega - \omega' + i\delta} \tag{1.2.31}$$

where

$$C_{AB}(\omega) = C^{+}_{AB}(\omega) + C^{-}_{AB}(\omega) \tag{1.2.32}$$

is the associated <u>spectral function</u>.

In particular we write in the case (1.2.6, 7)

$$C^{\pm}_{a_k a_k^+}(\omega) \equiv A^{\pm}(\omega) \tag{1.2.33}$$

and

$$C^{\pm}_{Q_q Q_q^+}(\omega) \equiv B^{\pm}(\omega). \tag{1.2.34}$$

By inspection of (1.2.27, 28) one finds the relations

$$A^{-}(k,\omega) = e^{-\beta\omega} A^{+}(k,\omega) \tag{1.2.35}$$

and

$$B^{-}(q,\omega) = e^{-\beta\omega} B^{+}(q,\omega) \tag{1.2.36}$$

so that

$$\left.\begin{aligned} A^{+}(k,\omega) &= (1 - f^{0}(\omega))\, A(k,\ ) \\ A^{-}(k,\omega) &= f^{0}(\omega)\, A(k,\omega) \end{aligned}\right\} \tag{1.2.37}$$

and

$$\left.\begin{aligned} B^{+}(q,\omega) &= (1 + n^{0}(\omega))\, B(q,\omega) \\ B^{-}(q,\omega) &= n^{0}(\omega)\, B(q,\omega). \end{aligned}\right\} \tag{1.2.38}$$

In the case of the phonons time reversal[9] implies in addition

$$\left.\begin{aligned} B^{-}(q,-\omega) &= -B^{+}(q,\omega) \\ B(q,-\omega) &= -B(q,\omega). \end{aligned}\right\} \tag{1.2.39}$$

The spectral function associated with the unperturbed functions (1.2.18), (1.2.19) are

$$A^{0}(k,\omega) = \delta(\omega - \epsilon^{o}_{k}) \tag{1.2.40}$$

$$B^o(q,\omega) = \tfrac{1}{2}(\delta(\omega-\omega^o_q) - \delta(\omega+\omega^o_q)) \quad (1.2.41)$$

From (1.2.27, 28, 33) the following sum rule is readily obtained

$$\int_{-\infty}^{+\infty} A(k,\omega)d\omega = \left\langle \left[a^+_k, a_k\right]_+ \right\rangle = 1. \quad (1.2.42)$$

The corresponding sum rule for $B(q,\omega)$ is trivially zero because of (1.2.39). In this case, however, one easily verifies the f - sum rule

$$\int_{-\infty}^{+\infty} B(q,\omega)\omega\, d\omega = \left\langle \left[Q_q \left[HQ^+_q\right]_-\right]_- \right\rangle = \omega^o_q \quad (1.2.43)$$

2.3 Perturbation Theory

In perturbation theory an expression like (1.2.1′) is calculated by a formal expansion in powers of the interaction, H_{int} in (1.2.14). This is obtained with the help of the evolution operator $S(t, t_o)$ defined by

$$\frac{\partial S(t,t_o)}{\partial t} = H_{int}[t]\,S(t,t_o); \quad S(t,t) = 1 \quad (1.2.44)$$

For time-independent Hamiltonians for which (1.2.2) and (1.2.16) hold we have in addition

$$S(t,t_o) = e^{iH_o t}\, e^{-iHt}\, e^{iHt_o}\, e^{-iH_o t_o} \quad (1.2.44')$$

By iteration one finds from (1.2.44)

$$S(t,t_o) = \sum_{n=0}^{\infty} S_n(t,t_o) \quad (1.2.45)$$

where the nth order term is given by

$$S_n(t, t_0) = \frac{1}{n!} T(-i \int_{t_o}^{t} dt' \; H_{int}[t'])^n \tag{1.2.45'}$$

and T is the time ordering operator.

With $S(t, t_0)$ the Heisenberg representation (1.2.2) may be reduced to the interaction representation (1.2.16)

$$A(t) = S(0, t) A[t] S(t, 0) \tag{1.2.46}$$

Because of the group property

$$S(t_1, t_2) S(t_2, t_3) = S(t_1, t_3) \tag{1.2.47}$$

Unitarity

$$S(t, t_2) = S^+(t_2, t_1) = S^{-1}(t_2, t_1) \tag{1.2.48}$$

and the time order of S the perturbation theoretic reduction has the general form

$$\begin{aligned}
&\langle T(A_1(t_1)\ldots A_m(t_m))\rangle \\
&= \mathrm{Tr}\left\{\rho T(S(0,t_1)A_1[t_1]S(t_1,t_2)A_2[t_2]\ldots A_m[t_m]S(t_m,0))\right\} \\
&= \mathrm{Tr}\left\{\rho S(0,+\infty) T(S(+\infty, -\infty)A_1[t_1]\ldots A_m[t_m])\right\} \\
&= \langle S^+(+\infty, -\infty) T(S(+\infty, -\infty) A_1[t_1]\ldots A_m[t_m]\rangle_o
\end{aligned} \tag{1.2.49}$$

where in the last step we made use of the switch-on condition

$$\rho[t] = S(t, -\infty) \rho^o S(-\infty, t) \tag{1.2.50}$$

and

$$\rho = \rho[0] \tag{1.2.50'}$$

Note that equations (1.2.50, 50′, 17′) imply (1.2.5′).

Equation (1.2.50) means that the interaction H_{int} was introduced adiabatically at $t = -\infty$. In the last line of equation (1.2.49) the power expansion in H_{int} is achieved since H_{int}, there, appears only in the S-matrix $S(+\infty, -\infty)$. Were it not for the factor $S^+(+\infty, -\infty)$ in front of the T-product Wick's theorem[10] could be applied to obtain the reduction into products of propagators (1.2.15) of the general n^{th} order term

$\langle T(S_n(+\infty, -\infty)\, A_1[t_1] \ldots A_m[t_m])\rangle_0$ which

apart from the vertex functions contained in the factors H_{int} (see equations (1.1.25, 30)) has the form $\langle T(A_1[t_1] \ldots A_N[t_N])\rangle_0$, $N \geqslant m$. This expression vanishes for odd N while for even N the reduction implied by Wick's theorem holds

$$\langle T(A_1[t_1] \ldots A_{2M}[t_{2M}])\rangle_o = \sum_{\text{all pairings }(k,\bar{k})} \prod_{s=1}^{M} \langle T(A_{k_s}[t_{k_s}]\, A_{\bar{k}_s}[t_{\bar{k}_s}])\rangle_0 \qquad (1.2.51)$$

Because of the factor $S^+(+\infty, -\infty)$ in (1.2.49) this procedure is applicable only at <u>zero temperature</u> where

$$\rho_{T=0} = |\Psi_0\rangle\langle\Psi_0| \;;\; \rho^0_{T=0} = |\phi_0\rangle\langle\phi_0|$$

$|\Psi_0\rangle$ and $|\phi_0\rangle$ being the ground states of H and H_0, respectively. If these ground states are non-degenerate then

$$S(+\infty, -\infty)\,|\phi_0\rangle = e^{i\alpha}|\phi_0\rangle$$

being a c-number and (1.2.49) factorizes

$$\langle\Psi_0|T(A_1(t_1)\ldots A_m(t_m))|\Psi_0\rangle$$

$$= \frac{\langle\phi_0|T(S(+\infty, -\infty)A_1[t_1]\ldots A_m[t_m])|\phi_0\rangle}{\langle\phi_0|S(+\infty, -\infty)|\phi_0\rangle} \quad (1.2.52)$$

This is the principal disadvantage of the simple real-time Green functions. For then perturbation theory only works at $T = 0$. We do not, therefore, discuss further this version of perturbation theory. Starting from equation (1.2.52) it proceeds in the same way as the version described in section 3.3.

For $T > 0$ there are essentially two alternative procedures: the imaginary-time Green functions and the double real-time Green functions.

3. IMAGINARY-TIME GREEN FUNCTIONS[3,7,11]

3.1 Definitions

The definitions analogous to (1.2.1, 1′) but with imaginary time arguments, $t = -i\tau$, in the Heisenberg representation (1.2.2) and with a time ordering operator $\mathcal{T}$ analogous to (1.2.3) but acting on the real variables τ (ordering along the imaginary time axes) are

$$\mathcal{G}_{AB}(\tau-\tau') = \langle\mathcal{T}(A(-i\tau)\,B(-i\tau'))\rangle \quad (1.3.1)$$

and

$$\mathcal{G}_{A_1\ldots A_m}(\tau_1-\tau_m \;\ldots\; \tau_{m-1}-\tau_m)$$
$$= \langle\mathcal{T}(A_1(-i\tau_1)\ldots A_m(-i\tau_m))\rangle . \quad (1.3.1')$$

They provide a powerful formalism at finite temperatures. The unperturbed Green function associated with (1.3.1) is

$$\mathcal{G}^{o}_{AB}(\tau-\tau') = \langle \mathcal{T}(A[-i\tau]B[-i\tau']) \rangle_o . \quad (1.3.2)$$

From the cyclicity of the trace,

$$\mathrm{Tr}(A_1A_2 \ldots A_m) = \mathrm{Tr}(A_2 \ldots A_m A_1)$$

follows the important periodicity

$$\mathcal{G}_{AB}(\tau) = \pm\mathcal{G}_{AB}(\tau-\beta) ; \quad 0<\tau<\beta \quad (1.3.3)$$

(+/- for boson/fermion operators) which allows one to write this function as a Fourier series

$$\mathcal{G}_{AB}(\tau) = \sum_{\sigma} \mathcal{G}_{AB}(-i\sigma)e^{i\sigma\tau} \quad (1.3.4)$$

Equation (1.3.3) implies

$$e^{i\sigma\beta} = \begin{cases} +1 ; & \text{bosons} \\ -1 ; & \text{fermions} \end{cases} \quad (1.3.5)$$

or

$$\sigma = \begin{cases} 2n\pi/\beta ; & \text{bosons} \\ (2n+1)\pi/\beta ; & \text{fermions} \end{cases} \quad (1.3.5')$$

with $n = 0, \pm 1, \pm 2, \ldots$. The inverse of (1.3.4) is

$$\mathcal{G}_{AB}(-i\sigma) = \beta^{-1}\int_0^{\beta} \mathcal{G}_{AB}(\tau)e^{-i\sigma\tau}\,d\tau . \quad (1.3.6)$$

The one-electron and one-phonon functions analogous to (1.2.6, 7) are

$$\mathcal{G}^{()}_{a_k a^+_{k'}}(\tau) \equiv \mathcal{G}^{()}(k,\tau)\,\delta_{\vec{k}\,\vec{k}'} \quad (1.3.7)$$

and

$$\mathcal{G}^{()}_{Q_q Q^+_{q'}}(\tau) \equiv \mathcal{D}^{()}(q,\tau)\,\delta_{\vec{q}\,\vec{q}'} . \quad (1.3.8)$$

The corresponding Fourier transformed unperturbed time-ordered functions are found to be

$$\mathcal{G}^{o}(k,-i\sigma) = \frac{\beta^{-1}}{\varepsilon^{o}_{k} + i\sigma} \quad (1.3.9)$$

and

$$\mathcal{D}^{o}(q, -i\sigma) = \frac{\beta^{-1}\omega_q^o}{\omega_q^{o\,2} + \sigma^2} \quad (1.3.10)$$

Making an analytic continuation into the complex frequency plane, $-i\sigma \rightarrow z$, we write the Dyson equations analogous to (1.2.22, 23) in the form

$$\mathcal{G}(k,z) = \mathcal{G}^{o}(k,z) + \beta\mathcal{G}^{o}(k,z)\Sigma(k,z)\mathcal{G}(k,z) \quad (1.3.11)$$

$$\mathcal{D}(q,z) = \mathcal{D}^{o}(q,z) - \beta\mathcal{D}^{o}(q,z)\Pi(q,z)\mathcal{D}(q,z) \quad (1.3.12)$$

so that, in analogy to (1.2.24, 25),

$$\mathcal{G}(k,z) = \frac{-\beta^{-1}}{z - \varepsilon_k^o - \Sigma(k,z)} \quad (1.3.13)$$

and

$$\mathcal{D}(q,z) = \frac{-\beta^{-1}\omega_q^o}{z^2 - \omega_q^{o2} - \omega_q^o\Pi(q,z)} \quad (1.3.14)$$

3.2 <u>Spectral Representation and Linear Response Theory</u>

From (1.3.1) and (1.2.8, 9) it follows that

$$\mathcal{G}_{AB}(\tau) = \Theta(\tau)iG_{AB}^{>}(-i\tau) + \Theta(-\tau)iG_{AB}^{<}(-i\tau) \quad (1.3.15)$$

which leads to the spectral representation of the analytically continued Fourier Transform (1.3.6)

$$\mathcal{G}_{AB}(z) = \beta^{-1}\int_{-\infty}^{+\infty} \frac{C_{AB}(\omega')d\omega'}{\omega' - z} \quad (1.3.16)$$

Comparison with (1.2.31) yields the result that the retarded one-particle Green function is the boundary value of the imaginary-time Green function,

$$G_{AB}^{ret}(\omega) = -\beta\mathcal{G}_{AB}(\omega + i\delta) \quad (1.3.17)$$

This relation is of great practical importance for the following reason: in any experiment with a many body system described by a hamiltonian H one acts on it via an external perturbation (which in general is time dependent), e.g. an electromagnetic field, a particle flux, a pressure, an ultrasonic wave, a thermal gradient or a heat pulse. In general this perturbation couples to some density operator (particle density, energy density, momentum density, etc.) of the system as described by

$$d_i(\vec{r}) = \sum_k \alpha_{ik} n_k(\vec{r}) \tag{1.3.18}$$

where $n_k(\vec{r})$ is the spectral density as given through its Fourier transform,

$$n_k(\vec{p}) = a^+_{\vec{k},s} \, a_{\vec{k}+\vec{p},s} \tag{1.3.19}$$

for electrons, or (see equation (1.1.23))

$$n_q(\vec{p}) = b^+_{\vec{q},\mu} \, b_{\vec{q}+\vec{p},\mu} \tag{1.3.20}$$

for phonons. For conserved particles (electrons) this is the only way to couple to the system. For non-conserved particles (phonons) one may also couple directly to the displacement field $\vec{u}_{Ib}$, (1.1.20). In this case, which occurs in the perturbation by an ultrasonic wave in a crystal,

$$d_i(\vec{r}) = \sum_{Ib} \vec{u}_{Ibi} \delta(\vec{r} - \vec{R}_{Ib}) \tag{1.3.21}$$

The coupling of external perturbations $\phi_i(\vec{r}, t)$ to these densities is then described by a hamiltonian

$$\begin{aligned} H'_t &= \int_V d^3r \sum_i \phi_i(\vec{r},t)\, d_i(\vec{r}) \\ &= V^{-1} \sum_{p,i} \phi_i(\vec{p}, t)\, d_i(-\vec{p}) \end{aligned} \tag{1.3.22}$$

The time evolution of the ensemble perturbed by H'_t is determined by the Liouville equation for the perturbed density matrix $\rho + \rho'_t$,

$$i \frac{\partial \rho'_t}{\partial t} = \left[H + H'_t, \rho + \rho'_t \right]_- \tag{1.3.23}$$

where ρ is the equilibrium density matrix (1.2.4). Assuming that the perturbation is adiabatically introduced at $t = -\infty$,

$$\rho'_{-\infty} = 0 \tag{1.3.24}$$

the solution of (1.3.23) is, to first order in H'_t,

$$\rho'_t = i \left[\rho, \int_0^\infty dt' e^{-\delta . t'} H'_{t-t'} (-t') \right]_- \tag{1.3.25}$$

Introducing the density variation

$$\delta d_i(\vec{r}) \equiv d_i(\vec{r}) - \langle d_i(\vec{r}) \rangle \tag{1.3.26}$$

and the perturbed average

$$\langle A \rangle_t \equiv \mathrm{Tr}\{ (\rho + \rho'_t) A \} \tag{1.3.27}$$

we have from (1.3.25, 22) (using the cyclicity of the trace)

$$\langle \delta d_i(\vec{r}) \rangle_t = \mathrm{Tr}[\rho'_t \, d_i(\vec{r})]$$

$$= \int_0^\infty dt' \; e^{-\delta . t'} \sum_{i'} \int d^3 r' \phi_{i'} \; (\vec{r}', t-t') \; .$$

$$. (-i) \langle \mathcal{R} (d_i(\vec{r}, t') \; d_{i'} \; (\vec{r}', 0)) \rangle \tag{1.3.28}$$

which is called a <u>Kubo formula</u>[12].

The last factor on the right of (1.3.28) which is recognized to be $G^{ret}_{d_i(\vec{r}) d_{i'}(\vec{r}')}(t')$, as defined by equation (1.2.10), is called a <u>dissipation function</u>.

It is in general not a one-particle but rather a two-particle Green function, since according to (1.3.18 - 20), and with the exception of (1.3.21), $d_i(\vec{r})$ is a two-particle operator.

Now since in the derivation of the spectral representations (1.2.31) and (1.3.16) it was not assumed that A, B were one-particle operators, such representations also hold for the dissipation function $G^{ret}_{d_i(\vec{r})d_{i'}(\vec{r}')}(t)$ and the imaginary-time Green function $\mathcal{G}_{d_i(\vec{r})d_{i'}(\vec{r}')}(\tau)$.

Therefore in calculating the latter, e.g. by perturbation theory, the dissipation function is obtained as boundary value according to (1.3.17).

3.3 Perturbation Theory

For imaginary times we again have an evolution operator analogous to (1.2.44)

$$\mathcal{S}(\tau,\tau_0) = S(-i\tau, -i\tau_0) = \sum_{n=0}^{\infty} \mathcal{S}_n(\tau,\tau_0) \quad (1.3.29)$$

where

$$\mathcal{S}_n(\tau,\tau_0) = \frac{1}{n!}\mathcal{T}\left(-\int_{\tau_0}^{\tau} d\tau'\, H_{int}[-i\tau']\right)^n \quad (1.3.30)$$

which gives the reduction of the Heisenberg representation to the interaction representations,

$$A(-i\tau) = \mathcal{S}(0,\tau)A[-i\tau]\mathcal{S}(\tau, 0) \quad (1.3.31)$$

In addition it also gives a similar reduction of the density matrix (1.2.4) to (1.2.17),

$$\rho = e^{\beta(\Omega-\Omega_0)}\rho^0\,\mathcal{S}(\beta, 0)\,. \quad (1.3.32)$$

In other words, the switch-on condition (1.2.50) which

gave rise to the difficulty of the factor $S^+(+\infty, -\infty)$ in the last expression of (1.2.50) is not needed, and one obtains for the analogue of (1.2.49)

$$\begin{aligned}
&\langle \mathcal{T}(A_1(-i\tau_1) \ldots A_m(-i\tau_m))\rangle \\
&= e^{\beta(\Omega - \Omega_0)} \mathrm{Tr}\{ \rho^0 \mathcal{S}(\beta,0) \mathcal{T}(\mathcal{S}(0, \tau_1) A_1[-i\tau_1] \\
&\quad \cdot \mathcal{S}(\tau_1, \tau_2) A_2[-i\tau_2] \ldots A_m [-i\tau_m] \mathcal{S}(\tau_m, 0))\} \\
&= e^{\beta(\Omega - \Omega_0)} \langle \mathcal{T}(\mathcal{S}(\beta, 0) A_1[-i\tau_1] \ldots A_m[-i\tau_m])\rangle_0
\end{aligned} \tag{1.3.33}$$

Here in the last step use was made of the time order of $\mathcal{S}(\tau, \tau_0)$, of the group property

$$\mathcal{S}(\tau_1, \tau_2)\mathcal{S}(\tau_2, \tau_3) = \mathcal{S}(\tau_1, \tau_3) \tag{1.3.34}$$

and of the fact that according to (1.3.3) all τ_i may be confined to the interval $0 < \tau_i < \beta$.

In the special case where the operator product in (1.3.33) is empty, $m = 0$, we find

$$e^{-\beta(\Omega - \Omega_0)} = \langle \mathcal{S}(\beta, 0)\rangle_0 \tag{1.3.35}$$

so that, in analogy to (1.2.52),

$$\langle \mathcal{T}(A_1(-i\tau_1) \ldots A_m(-i\tau_m))\rangle = \frac{\langle \mathcal{T}(\mathcal{S}(\beta, 0) A_1[-i\tau_1] \ldots A_m[-i\tau_m])\rangle_0}{\langle \mathcal{S}(\beta, 0)\rangle_0} \tag{1.3.36}$$

It is important that in this formalism again a "Wick's theorem" reduction[13,14] analogous to (1.2.51) holds for an even number of factors (the expression vanishes for an odd number of factors)

$$\langle \mathcal{T}(A_1 [-i\tau_1] \ldots A_{2M}[-i\tau_{2M}])\rangle_o =$$

$$= \sum_{\text{all pairings } (k,\, k)} \prod_{s=1}^{M} \langle \mathcal{T}(A_{k_s}[-i\tau_{k_s}]\, A_{\bar{k}_s}\, [-i\tau_{\bar{k}_s}]\,)\rangle_o \quad (1.3.37)$$

Thus the nth order term

$$\langle \mathcal{T}(\mathcal{S}_n(\beta,\, 0)A_1\, [-i\tau_1] \ldots A_m\, [-i\tau_m])\rangle_o$$

in the numerator of (1.3.36) is resolved into a sum of products of propagator lines $g^{o}_{AB}(\tau)$ (= B——A) connected through the vertex functions

$V_{kk'\ldots}$ (= 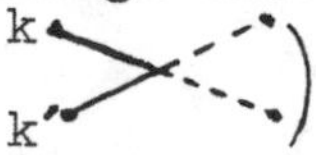) contained in the factors

$$H_{int} = \sum_{kk'\ldots} V_{kk'\ldots}\, A_k\, A_{k'}\ldots \quad (1.3.38)$$

The terms of this sum for

$$\langle \mathcal{T}(\mathcal{S}_n\, (\beta,\, 0)A_1\, [-i\tau_1] \ldots A_m\, [-i\tau_m])\rangle_o$$

are called n^{th} order <u>Feynman diagrams</u> with m external lines.

As is seen from (1.3.35) the diagrams with no external lines $(m = 0)$ describe the thermal equilibrium variation $\Omega - \Omega_o$ of the thermodynamic potential as due to the dynamic interaction H_{int}. We call these <u>equilibrium diagrams</u>. They are also of importance for the diagrams with external lines in that the latter may be disconnected in the sense as to contain equilibrium diagrams as factors. Calling a diagram <u>connected</u> if it contains no equilibrium diagrams as factors one finds that in (1.3.36) these factors just cancel the denominator[14], so that

$$\left\langle \mathcal{T}(A_1(-i\tau_1) \ldots A_m(-i\tau_m))\right\rangle =$$
$$= \left\langle \mathcal{T}(\mathcal{S}(\beta,0)A_1[-i\tau_1] \ldots A_m[-i\tau_m])\right\rangle_{0,\ conn.} \tag{1.3.39}$$

(The same conclusion is also true for (1.2.52)).

The definition of connectedness also applies to an equilibrium diagram. In this case one can show that[14]

$$\langle \mathcal{S}(\tau, 0)\rangle_0 = \exp\langle \mathcal{S}(\tau,0) - 1\rangle_{0,\ conn.} \tag{1.3.40}$$

or with (1.3.35)

$$\Omega - \Omega_0 = \beta^{-1}\langle \mathcal{S}(\beta,0) - 1\rangle_{0,\ conn.} \tag{1.3.41}$$

4. DOUBLE REAL-TIME GREEN FUNCTIONS

4.1 Equilibrium Green Functions

This is a more recent development of the theory, due essentially to Keldysh[15], which allows Green functions and dissipation functions at finite temperature to be calculated without use of the unphysical imaginary times. The basic idea is that a time ordering T_C may be defined along any path C in the complex plane, not only along the real or imaginary axis, and that it is possible to choose a path C such that all the factors $S(t, t')$ in (1.2.49) can be absorbed in the time-ordered product.

To this end we insert (1.2.50, 50′) into the

second line of (1.2.49) and use the cyclicity of the trace, so that

$$\langle T(A_1(t_1)\,\ldots\,A_m(t_m))\rangle =$$
$$= \langle S(-\infty, 0)\; T(S(0,t_1)A_1[t_1]\; S(t_1,t_2)\; A_2[t_2]\ldots$$
$$\ldots A_m[t_m]\, S_m(t_m, 0))\; S(0, -\infty)\rangle_0 \qquad (1.4.1)$$

Now we choose the path C such as to run through the real time axis twice, first in the direction of increasing values (called t_+) and then in the direction of decreasing values (called t_-), as shown in fig. 1 below.

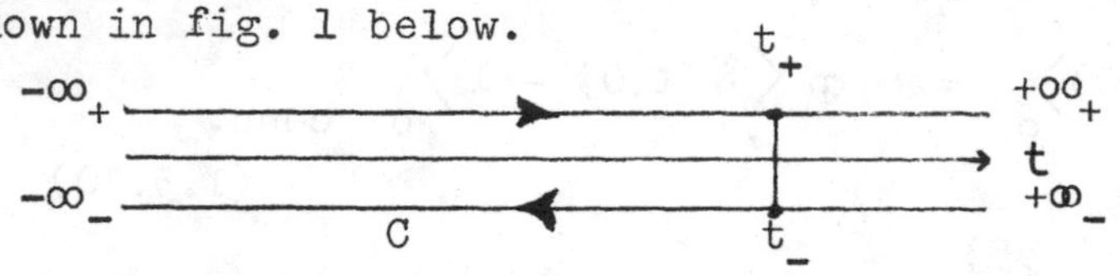

Figure 1.

The time ordering operator T_C associated with this path C is given by (1.2.3) but with the step functions $\Theta(t-t')$ replaced by

$$\Theta(t, t') = \left\{\begin{array}{ll} \Theta(t-t') ; & t = t_+,\ t' = t'_+ \\ 1 ; & t = t_-,\ t' = t'_+ \\ 0 ; & t = t_+,\ t' = t'_- \\ \Theta(t'-t) ; & t = t_-,\ t' = t'_- \end{array}\right\} \qquad (1.4.2)$$

with the definition

$$S_C(t, t_0) = T_C \exp\left(-i \int_{C\,t_0}^{t} dt' H_{int}[t']\right) \qquad (1.4.3)$$

and replacing (1.2.50) by

$$\rho[t] = S_C(t, -\infty_+)\, \rho^0\, S_C(-\infty_-, t) / \langle S_C\rangle_0 \qquad (1.4.4)$$

where the denominator insures that

$$\mathrm{Tr}\,\rho[t] = 1 \qquad (1.4.4')$$

and

$$S_C \equiv S_C(-\infty_-, -\infty_+) \qquad (1.5.5)$$

we have, in slight generalization of (1.4.1),

$$\langle T_C(A_1(t_1)\ldots A_m(t_m))\rangle = \frac{\langle T_C(S_C\, A_1[t_1]\ldots A_m[t_m])\rangle_0}{\langle S_C\rangle_0} \qquad (1.4.6)$$

Since the numerator of (1.4.6) is now completely time-ordered we can apply Wick's theorem which, as in (1.2.51), leads to a reduction into propagators

$$\langle T_C(A_1[t_1]\ldots A_{2M}[t_{2M}])\rangle_0 = \sum_{\text{all pairings }(k\bar{k})} \prod_{s=1}^{M} \langle T_C(A_{k_s}[t_{k_s}]\, A_{\bar{k}_s}[t_{\bar{k}_s}])\rangle_0 \qquad (1.4.7)$$

In addition, the disconnected parts in the numerator of (1.4.6) factor and cancel the denominator as in (1.2.52) and in (1.3.36), so that

$$\langle T_C(A_1(t_1)\ldots A_m(t_m))\rangle = \langle T_C(S_C\, A_1[t_1]\ldots A_m[t_m])\rangle_{0,\mathrm{conn.}} \qquad (1.4.8)$$

The unperturbed Green functions appearing on the right of (1.4.7) are

$$\begin{aligned}
\langle T_C(A[t_+]B[t'_+])\rangle_0 &= \langle T(A[t]B[t'])\rangle_0 &&= iG^0_{AB}(t-t')\\
\langle T_C(A[t_-]B[t'_+])\rangle_0 &= \langle A[t]B[t']\rangle_0 &&= iG^{0>}_{AB}(t-t')\\
\langle T_C(A[t_+]B[t'_-])\rangle_0 &= \pm\langle B[t']A[t]\rangle_0 &&= iG^{0<}_{AB}(t-t')\\
\langle T_C(A[t_-]B[t'_-])\rangle_0 &= \langle \overline{T}(A[t]B[t'])\rangle_0 &&= i\overline{G}^0_{AB}(t-t')
\end{aligned} \qquad (1.4.9)$$

where $\overline{T}$ and $\overline{G}^o_{AB}$ refer to the anti-time order. The full Green functions are obtained with the help of (1.4.8)

$$\langle T_C(S_C\, A[t_+]B[t'_+])\rangle_{o,conn.} = \langle T(A(t)B(t'))\rangle = iG_{AB}(t-t')$$
$$\langle T_C(S_C\, A[t_-]B[t'_+])\rangle_{o,conn.} = \langle A(t)B(t')\rangle = iG^{>}_{AB}(t-t')$$
$$\langle T_C(S_C\, A[t_+]B[t'_-])\rangle_{o,conn.} = \pm\langle B(t')A(t)\rangle = iG^{<}_{AB}(t-t')$$
$$\langle T_C(S_C\, A[t_-]B[t'_-])\rangle_{o,conn.} = \langle \overline{T}(A(t)B(t'))\rangle = i\overline{G}_{AB}(t-t') \quad (1.4.10)$$

The remarkable feature about this formalism is that it contains all the different one-particle Green functions in one single definition.

4.2 Non-Equilibrium Green Functions

An external perturbation of the form (1.3.22) gives rise to a non-equilibrium situation which in Section 3 was described by the Kubo formula (1.3.28). An alternative and more general formulation of non-equilibrium quantities is obtained with the technique of functional derivatives[8, 16]. In the form of double real-time Green functions it is due to Niklasson and Sjölander [17]. The evolution operator describing the perturbation H'_t in the double time formulation is given by

$$\mathbb{S}_C(t,t_o) = T_C \exp(-i\int_{t_o\,C}^{t} d1\, \phi(1)d(1)) \quad (1.4.11)$$

where $\phi(1) \equiv \phi_{i_1}(\vec{r}_1, t_1)$, $d(1) \equiv d_{i_1}(\vec{r}_1, t_1)$ (in the Heisenberg representation) and integration is over $\vec{r}_1$ and t_1 (along C) and includes summation over

i_1. If in analogy to (1.2.46) the time dependence of $d_i(\vec{r})$ under the combined action of $H + H'_t$ is defined by

$$d_\emptyset(1) \equiv \mathbb{S}_C(0,t_1)d(1)\mathbb{S}_C(t_1, 0) \qquad (1.4.12)$$

and in analogy to (1.4.4), (1.2.50′)

$$\rho_\emptyset = \mathbb{S}_C(0, -\infty_+)\rho\,\mathbb{S}_C(-\infty_-, 0)/\langle\mathbb{S}_C\rangle \quad (1.4.13)$$

then the analogue of (1.4.6) holds and defines the non-equilibrium m-particle Green function $\mathbb{G}_m$

$$\begin{aligned}(-i)^{-m/2}\mathbb{G}_m(1,\ldots m) &\equiv \\ &\equiv \mathrm{Tr}\left\{\rho_\emptyset\, T_C\,(d_\emptyset(1)\,\ldots\,d_\emptyset(m))\right\} \equiv \\ &\equiv \langle T_C(d_\emptyset(1)\,\ldots\,d_\emptyset(m))\rangle_\emptyset = \\ &= \langle T_C(\mathbb{S}_C\,d(1)\,\ldots\,d(m))\rangle/\langle\mathbb{S}_C\rangle \qquad (1.4.14)\end{aligned}$$

where

$$\mathbb{S}_C \equiv \mathbb{S}_C(-\infty_-, -\infty_+)\,. \qquad (1.4.15)$$

An alternative definition of the Green function $\mathbb{G}_m$ is obtained by considering $\langle\mathbb{S}_C\rangle$ as its generating functional. Then

$$\mathbb{G}_m(1,\ldots m) = \frac{(-i)^{-m/2}}{\langle\mathbb{S}_C\rangle}\cdot\frac{\delta^m\langle\mathbb{S}_C\rangle}{\delta\phi(1)\,\ldots\,\delta\phi(m)} \qquad (1.4.16)$$

as is easily verified with the help of (1.4.11, 14, 15). In the non-equilibrium case the associated <u>irreducible parts</u> or <u>cumulants</u> are of importance. They are generated from the functional $\log\langle\mathbb{S}_C\rangle = \langle\mathbb{S}_C - 1\rangle_{irr}$

$$\mathbb{G}_m^{irr}(1,\ldots,) = (-i)^{-m/2}\,\frac{\delta^m\log\langle\mathbb{S}_C\rangle}{\delta\phi(1)\,\ldots\,\delta\phi(m)} \qquad (1.4.17)$$

and have the important recurrence property

$$\mathbb{G}_m^{irr}(1,\ldots m) = (-i)^{\frac{n-m}{2}} \frac{\delta^{m-n}\,\mathbb{G}_n^{irr}(1,\ldots n)}{\delta\phi(n+1)\ldots\delta\phi(m)} \tag{1.4.18}$$

valid for all $n \leqslant m$. The first two cumulants are

$$\mathbb{G}_1^{irr}(1) = \mathbb{G}_1(1) = (-i)^{\frac{1}{2}} \langle d_\phi(1) \rangle_\phi \tag{1.4.19}$$

and

$$\mathbb{G}_2^{irr}(1,2) = \mathbb{G}_2^{irr}(2,1) = \frac{\delta\langle d_\phi(1)\rangle_\phi}{\delta\phi(2)} =$$

$$= \mathbb{G}_2(1,2) - \mathbb{G}_1(1)\mathbb{G}_1(2) =$$

$$= -i\,\langle T_C(\Delta d_\phi(1)\,\Delta d_\phi(2))\rangle_\phi \tag{1.4.20}$$

where

$$\Delta d_\phi(1) = d_\phi(1) - \langle d_\phi(1)\rangle_\phi \tag{1.4.21}$$

Equation (1.4.20) may also be written as

$$\delta\langle d_\phi(1)\rangle_\phi = -i\int_C d2\,\langle T_C(\Delta d_\phi(1)\Delta d_\phi(2))\rangle_\phi\,\delta\phi(2)$$

or by decomposition of the path C into its ascending ($t = t_+$) and descending ($t - t_-$) parts and using the notation $1_\pm \equiv (i_1, \vec{r}_1, t_{1\pm})$,

$$\delta\langle d_\phi(1_+)\rangle_\phi = -i\int_{-\infty}^{t_1} d2\,\Big\{\langle\Delta d_\phi(1)\Delta d_\phi(2)\rangle_\phi\,\delta\phi(2_+)$$
$$- \langle\Delta d_\phi(2)\,\Delta d_\phi(1)\rangle_\phi\;\delta\phi(2_-)\Big\} -$$
$$-i\int_{t_1}^{+\infty} d2\,\langle\Delta d_\phi(2)\,\Delta d_\phi(1)\rangle_\phi\{\delta\phi(2_+)-\delta\phi(2_-)\} \tag{1.4.22}$$

While in (1.4.22) the variation $\delta\phi(2)$ was arbitrary along the whole path C the following particular variations are also of interest:

$$\delta\phi(2_+) \equiv \delta\phi(2_-) \equiv \delta_R\,\phi(2) \qquad (1.4.23)$$

for which (1.4.22) becomes

$$\delta\langle d_\phi(1_+)\rangle_\phi = -i\int_{-\infty}^{t_1} d2 \left\langle \left[d_\phi(1),\, d_\phi(2)\right]_-\right\rangle_\phi \delta_R\,\phi(2)$$

which is a generalization of the Kubo formula (1.3.28), or

$$\frac{\delta\langle d_\phi(1_+)\rangle_\phi}{\delta_R\,\phi(2)} = -i\langle \mathcal{R}\,(d_\phi(1)\, d_\phi(2))\rangle_\phi$$

$$\equiv \mathbb{G}^{ret}\,(1,2) \qquad (1.4.24)$$

where $\mathcal{R}$ is the retardation operator (1.2.11). This defines the non-equilibrium dissipation function which, for $\phi(1) \equiv 0$, reduces to the dissipation function $G^{ret}_{d_{i_1}(\vec{r}_1)d_{i_2}(\vec{r}_2)}\,(t_1 - t_2)$ of the Kubo formula (1.3.28).

If, on the other hand, we put

$$\delta\phi(2_+) \equiv 0, \quad \delta\phi(2_-) \equiv -\,\delta_<\phi(2) \quad (1.4.25)$$

(1.4.22) yields

$$\delta\langle d_\phi(1_+)\rangle_\phi = -i\int_{-\infty}^{+\infty} d2\langle \Delta\, d_\phi(2)\,\Delta d_\phi(1)\rangle_\phi \,.\,\delta_<\phi(2)$$

or

$$\frac{\delta\langle d_\phi(1_+)\rangle_\phi}{\delta_<\phi(2)} = -i\,\langle \Delta\, d_\phi(2)\,\Delta d_\phi(1)\,\rangle_\phi$$

$$\equiv \mathbb{G}^<(1,2) = \mathbb{G}^>(2,1) \quad . \qquad (1.4.26)$$

which for $\phi(1) \equiv 0$ reduce to the correlation functions

$$G^<_{d_{i_1}(\vec{r}_1)d_{i_2}(\vec{r}_2)}\,(t_1-t_2) = G^>_{d_{i_2}(\vec{r}_2)d_1(\vec{r}_1)}(t_2-t_1)$$

defined in (1.2.8,9).

II. PHONON-PHONON INTERACTION IN DIELECTRIC CRYSTALS AND IN SUPERFLUIDS

1. ULTRASONIC ATTENUATION IN DIELECTRIC CRYSTALS

1.1 Linear Response to an Oscillating External Force[18)]

In the case of an ultrasonic wave in a crystal the "external" perturbation, $\phi_i(\vec{r},t)$ of Part I section 3.2, is an oscillating force $\vec{F}_{Ib}\, e^{-i\omega t}$ acting on the ions at $\vec{R}_{Ib}$, and the density is given by (1.3.21). The perturbing hamiltonian (1.3.22) is then

$$H'_t = \sum_x F_x\, u_x\, e^{-i\omega t} \tag{2.1.1}$$

where, as in Part I, section 1.1, $x \equiv (I,b,i)$. It follows from (1.1.20, 23) that $\langle u_x \rangle = 0$ so that with (1.2.13) the Kubo formula (1.3.28) simply reads

$$\langle u_x \rangle_t = e^{-i\omega t} \sum_{x'} G^{ret}_{u_x u_{x'}}(\omega) F_{x'} \tag{2.1.2}$$

Applying (1.1.20) and defining

$$\widetilde{F}(q) \equiv \sum_x \sqrt{\frac{N}{M_o\, \omega^o_q}}\, \psi_{-q}(x)\, F_x \tag{2.1.3}$$

we may write, making use of (1.2.7) (which is a matrix in μ, μ') and of $\pm q \equiv (\pm \vec{q}, \mu)$

$$\langle u_{Ib} \rangle_t = e^{-i\omega t} \frac{1}{N} \sum_q \sum_{\mu\mu'} \frac{\vec{e}_b(\vec{q},\mu)\, e^{i\vec{q}\cdot\vec{R}_I}}{\sqrt{M_b\, \omega^o_{\vec{q},\mu}}}\, D^{ret}_{\mu,\mu'}(\vec{q},\omega)\, \widetilde{F}(\vec{q},\mu') \tag{2.1.4}$$

The physical interpretation of (2.1.3) is that $\widetilde{F}(\vec{q},\mu)$ is the "external" displacement wave with vector $\vec{q}$ and polarization and mode index μ, projected on the polarization vector $\vec{e}_b(\vec{q},\mu)$, and averaged over the positions of the ions of the unit cell.

Experimentally the force $\vec{F}_{Ib}$ acting on the individual ions cannot, of course, be controlled. What one has under control is the polarization and mode μ and the direction of propagation $\hat{q} \equiv \vec{q}/|\vec{q}|$ of the imposed wave $\widetilde{F}(\vec{q},\mu)$. The magnitude $p \equiv |\vec{q}|$ of the wave vector, on the other hand, is determined by the response of the system. Therefore we can choose, going over to a continuous $\vec{q}$,

$$\widehat{F}(\vec{q},\mu) = \widetilde{F}(p)\,\delta(\hat{q}-\hat{q}_0)\,\delta_{\mu\mu_0} \qquad (2.1.5)$$

so that in using (I.3.17)

$$\langle u_{Ib}\rangle_t = e^{-i\omega t}\,\frac{V}{N}\int_0^\infty p^2 dp\,\widehat{F}(p)\sum_\mu \frac{\vec{e}_b(p.\hat{q}_0,\mu)e^{ip\hat{q}_0.\vec{R}_I}}{\sqrt{M_b\,\omega^0_{p.\hat{q}_0,\mu}}}\,.$$

$$.(-\beta)\,\mathcal{D}_{\mu\mu_0}(p.\hat{q}_0,\omega+i\delta)\,. \qquad (2.1.6)$$

Writing the matrix $\mathcal{D}$ in the form (1.3.14) one may, to lowest order, retain only the diagonal part of the self-energy matrix Π and treat the non-diagonal part as perturbation. Now the denominator of (1.3.14) has in general quasi-particle poles in the unphysical sheet of the complex z-plane for fixed q, corresponding to first and second sound[19,20]. Because of

$$\left.\begin{aligned}\mathcal{D}(q,z) &= \mathcal{D}^*(q,z^*) = \mathcal{D}(q,-z)\\ \Pi(q,z) &= \Pi^*(q,z^*) = \Pi(q,-z)\end{aligned}\right\}\qquad (2.1.7)$$

which follows from (1.3.16, 8, 14), (1.2.39,32,34,27,28)

these poles always occur simultaneously at $\pm z$ and $\pm z^*$. Retaining only the diagonal part of Π the position of the poles

$$z = \pm(\omega_q \pm i\Gamma_q) \; ; \; \omega_q \gg \Gamma_q > 0 \qquad (2.1.8)$$

is determined by

$$(\omega_q \pm i\Gamma_q)^2 - \omega_q^{o\,2} - \omega_q^o \Pi(q, \omega_q \pm i\Gamma_q) = 0 \qquad (2.1.9)$$

or for small renormalization $\Pi(q, \omega_q \pm i\Gamma_q)$

$$\left.\begin{aligned} \omega_q &\cong \omega_q^o + \tfrac{1}{2} \operatorname{Re}\Pi(q, \omega_q^o + i\delta) \\ \Gamma_q &\cong +\tfrac{1}{2} \operatorname{Im}\Pi(q, \omega_q^o + i\delta) \; . \end{aligned}\right\} \qquad (2.1.10)$$

ω_q is the renormalized frequency and Γ_q the width of the quasi-particle. The positions of the poles, taken as functions of complex p and with ω (real), q_o and μ_o fixed, will determine the response of (2.1.6). The pole with smallest $|\operatorname{Im} p|$ will dominate and determine the wavelength of the propagated sound wave. Writing

$$\omega_{p.\hat{q}_o, \mu_o} \equiv \omega(p) \; ; \; \Gamma_{p.\hat{q}_o, \mu_o} \equiv \Gamma(p) \qquad (2.1.11)$$

the positions $p = p_o + i\alpha$ of the poles, as determined by (2.1.8) with $z = \omega + i\delta$, are obtained from

$$\pm(\omega(p) \pm i\Gamma(p)) = \omega \; . \qquad (2.1.12)$$

For $|\alpha/p_o| \ll 1$ only the upper external sign is possible and we may write

$$\omega(p_o + i\alpha) = \omega(p_o) + i\alpha V(p_o) \qquad (2.1.13)$$

where

$$V(p) \equiv \hat{q}_o \vec{V}_{p.\hat{q}_o} \qquad (2.1.14)$$

and $\vec{V}_q \equiv \partial\omega_q/\partial\vec{q}$ is the group velocity. Equation (2.2.13) inserted in (2.2.12) gives to lowest order

$$\omega(p_0) = \omega, \quad \text{or} \quad p_0 = p_0(\omega, \hat{q}_0, \mu_0) \tag{2.1.15}$$

and

$$\alpha = \frac{\Gamma(p_0)}{V(p_0)} = \alpha(\omega, \hat{q}_0, \mu_0) \tag{2.1.16}$$

Now (2.1.6) may be calculated approximately by contour integration around the dominating pole $p_0 + i\alpha$. Writing $\vec{q}_0 \equiv p_0\,\hat{q}_0$, $q_0 \equiv (\vec{q}_0, \mu_0)$ we find

$$\langle \vec{u}_{Ib} \rangle_t \cong Z(q_0)\, e^{i(\vec{q}_0\cdot\vec{R}_I - \omega t)}\, e^{-\alpha\hat{q}_0\cdot\vec{R}_I} \tag{2.1.17}$$

where

$$Z(q_0) = \\ = \frac{V}{N}\, p_0^2\, \tilde{F}(p_0)\vec{e}_b(\vec{q}_o)\sqrt{\omega^o_{q_o}/M_b}\;\frac{1}{2\omega}\oint_{p_0+i\alpha}\frac{dp}{\omega - \omega(p) + i\Gamma(p)} \tag{2.1.18}$$

Thus α is recognized as the <u>attenuation coefficient</u>. It is given by (2.1.16), or with (2.1.10,11,14) by

$$\alpha(q_o) \equiv \alpha(\omega^o_{q_o}, \hat{q}_o, \mu_o) = \frac{\operatorname{Im}\Pi(q_o, \omega^o_{q_o} + i\delta)}{2\hat{q}_0\cdot\vec{V}_{q_0}} \tag{2.1.19}$$

1.2 Attenuation of Acoustic Phonons

In dielectric crystals the self-energy Π is due to the anharmonic interaction (1.1.18,25). In the perturbation theory of Part I, section 3 Π is obtained by applying (1.3.39,37) to the calculation of $\mathcal{D}(q,\tau)$ and identifying Π through the

Dyson equation (1.3.12). Using the fact that according to (1.1.23) and (1.2.21)

$$\mathcal{D}^0(q, \tau = +0) = n_q^0 + \tfrac{1}{2} \qquad (2.1.20)$$

the first order term $\mathcal{S}_1(\beta, 0)$ yields

$$\Pi_1(q, -i\sigma) = \tfrac{1}{2} \sum_{q} C^{(4)}_{-q,q,q',-q'} (n^o_{q'} + \tfrac{1}{2}) \qquad (2.1.21)$$

which is real and therefore does not contribute to (2.1.19). The second order term $\mathcal{S}_2(\beta, 0)$ yields

$$\Pi_2(q, -i\sigma) =$$

$$= -\frac{\beta}{2} \sum_{q_1 q_2} \left| C^{(3)}_{-q\, q_1 q_2} \right|^2 \sum_{\sigma'} \mathcal{D}^0(q_1, -i\sigma') \mathcal{D}^0(q_2, -i\sigma + i\sigma') \qquad (2.1.22)$$

The diagrams corresponding to Π_1 and Π_2 are shown in fig. 2(a) and 2(b)

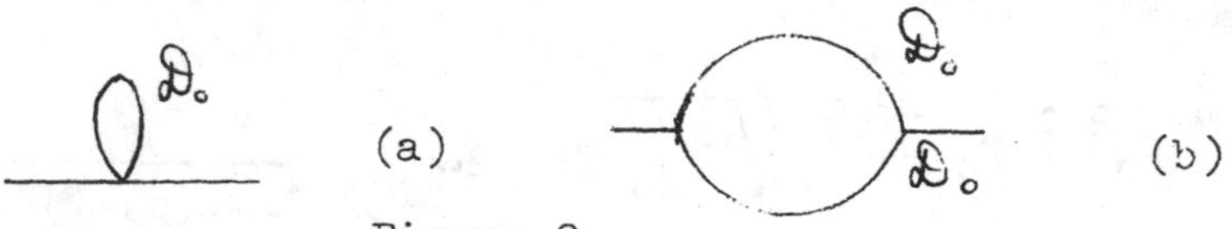

Figure 2.

To evaluate the σ'-sum we may use the identity

$$\pm \beta^{-1} \sum_{\sigma} f(i\sigma) = \frac{1}{2\pi i} \int_C \frac{f(z)dz}{e^{\beta z} \pm 1} \qquad (2.1.23)$$

(+/- for bosons/fermions, see (1.3.5′)) valid for any function $f(z)$ which is analytic in the neighbourhood of the imaginary z-axis. C is the path shown in fig. 3.

Equation (2.1.23) follows from the fact that according to (1.3.5) the denominator on the right has simple poles at the points $z = i\sigma$. Application

bosons: fermions:

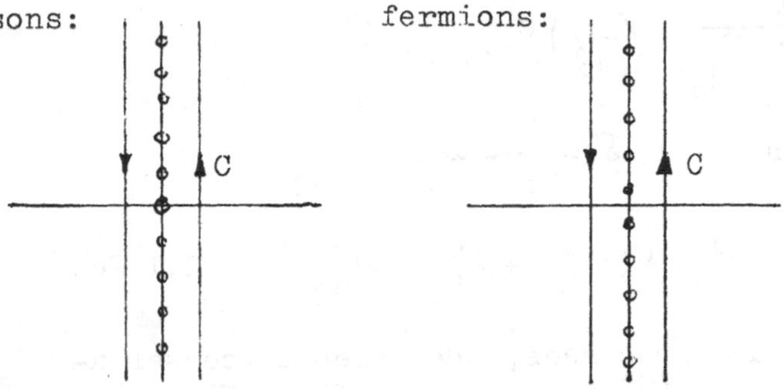

Figure 3.

of (2.1.23) yields the unperturbed "Lindhart function"

$$L^{0} \equiv 4\beta\sum_{\sigma'}\mathcal{D}^{0}(q_1, -i\sigma')\mathcal{D}^{0}(q_2, -i\sigma+i\sigma') =$$

$$= \frac{n^{0}_{q_1}+n^{0}_{q_2}+1}{i\sigma+\omega^{0}_{q_1}+\omega^{0}_{q_2}} + \frac{n^{0}_{q_1}+n^{0}_{q_2}+1}{-i\sigma+\omega^{0}_{q_1}+\omega^{0}_{q_2}}$$

$$+ \frac{n^{0}_{q_1}-n^{0}_{q_2}}{i\sigma-\omega^{0}_{q_1}+\omega^{0}_{q_2}} + \frac{n^{0}_{q_1}-n^{0}_{q_2}}{-i\sigma-\omega^{0}_{q_1}+\omega^{0}_{q_2}} \qquad (2.1.24)$$

and hence after analytic continuation $(-i\sigma\to\omega+i\delta)$

$$\mathrm{Im}\,\Pi_2(q,\omega+i\delta) = \frac{\pi}{8}\sum_{q_1q_2}\left|C^{(3)}_{-q\;q_1q_2}\right|^2 \cdot$$

$$\cdot\Big\{(n^{0}_{q_1}+n^{0}_{q_2}+1)\left[\delta(\omega-\omega^{0}_{q_1}-\omega^{0}_{q_2}) - \delta(\omega+\omega^{0}_{q_1}+\omega^{0}_{q_2})\right]$$

$$+ 2(n^{0}_{q_1}-n^{0}_{q_2})\,\delta(\omega+\omega^{0}_{q_1}-\omega^{0}_{q_2})\,. \qquad (2.1.25)$$

Hence

$$\alpha(q_0) = \frac{\pi}{16\hat{q}_0\cdot\vec{v}_{q_0}} \sum_{q_1q_2} |C^{(3)}_{-q_0q_1q_2}|^2 \cdot$$

$$\cdot\Big\{(n^o_{q_1} + n^o_{q_2} + 1)\,\delta(\omega^o_{q_o} - \omega^o_{q_1} - \omega^o_{q_2}) +$$

$$+ 2(n^o_{q_1} - n^o_{q_2})\,\delta(\omega^o_{q_o} + \omega^o_{q_1} - \omega^o_{q_2})\Big\}. \quad (2.1.26)$$

Neglecting Umklapp processes, the momentum conservation contained in $C^{(3)}_{-q_0q_1q_2}$ implies $\vec{q}_0 = \vec{q}_1 + \vec{q}_2$. The first and second terms in the bracket of (2.1.26) correspond respectively to the process $\mu_0 \rightleftharpoons \mu_1+\mu_2$ and $\mu_0+\mu_1 \rightleftharpoons \mu_2$. If μ_0 is a transverse acoustic phonon, $(\mu_0 = t_0)$, then energy and momentum conservation imply that the dominant contribution to (2.1.26) is the Landau-Rumer process $t_0 + \ell_1 \rightleftharpoons \ell_2$ where $\mu = \ell$ is a longitudinal acoustic phonon. For $\omega^o_q \ll T \ll T_D$, (T_D is the Debye temperature) and in the elastic continuum model where $C^{(3)}_{q_0q_1q_2}$ is of the form (1.1.51) and

$$\omega^o_{\vec{q}t} = c^o_t\,|\vec{q}|\;;\quad \omega^o_{\vec{q}\ell} = c^o_\ell\,|\vec{q}| \quad (2.1.27)$$

one finds $\alpha \propto \omega T^4$.

Under the same conditions the main contributions to (2.1.26) in the case $\mu_0 = \ell_0$ come from $\ell_0 \rightleftharpoons t_1+t_2$, $\ell_0 \leftrightharpoons t_1+\ell_2$ and $\ell_0+t_1 \rightleftharpoons \ell_2$ and lead to $\alpha \propto \omega^4 T$. However the order of magnitude turns out to be too small. One explanation is that, due to the finite lifetimes of the phonons, energy conservation is not strictly valid. Then the process $\ell_0+\ell_1 \rightleftharpoons \ell_2$

which in the approximation (2.1.27) is strictly <u>collinear</u> and therefore has a drastically reduced phase space, becomes important. In addition processes involving more than three phonons cannot be ignored.

The <u>lifetime effect</u> is taken care of by replacing in (2.1.22) the unperturbed Green functions $\mathcal{D}^0$ by the exact functions $\mathcal{D}$ and using the spectral representation (1.3.16), (1.2.32,34)

$$\Pi_2(q,-i\sigma) = -\frac{1}{2\beta}\sum_{q_1q_2}\left|C^{(3)}_{-q\,q_1q_2}\right|^2\sum_{\sigma'}\int\frac{d\omega_1 B(q_1,\omega_1)}{\omega_1+i\sigma'}\int\frac{d\omega_2 B(q_2,\omega_2)}{\omega_2+i\sigma-i\sigma'} \tag{2.1.28}$$

With the aid of (2.1.23) the σ'-sum reduces to

$$\beta^{-1}\sum_{\sigma'}\frac{1}{\omega_1+i\sigma'}\,\frac{1}{\omega_2+i\sigma-i\sigma'} = \frac{n^0(\omega_1)+n^0(\omega_2)+1}{\omega_1+\omega_2+i\sigma} \tag{2.1.29}$$

so that after analytic continuation $-i\sigma\rightarrow\omega+i\delta$

$$\mathrm{Im}\Pi_2(q,\omega+i\delta) = \frac{\pi}{2}\sum_{q_1q_2}\left|C^{(3)}_{-q\,q_1q_2}\right|^2\int d\omega_1\int d\omega_2\, B(q_1,\omega_1)\,B(q_2,\omega_2)\,. \\ (n_0(\omega_1)+n^0(\omega_2)+1)\,\delta(\omega-\omega_1-\omega_2) \tag{2.1.30}$$

For the unperturbed spectral function (1.2.41) this readily goes over into (2.1.25).

At this stage one usually approximates the spectral function by a Lorentzian shape

$$B(q,\omega) = \frac{\Gamma_q\,\omega_q^o}{2\pi\,\omega_q}\left\{\frac{1}{(\omega-\omega_q)^2+\Gamma_q^{\,2}}-\frac{1}{(\omega+\omega_q)^2+\Gamma_q^{\,2}}\right\} \tag{2.1.31}$$

Note that this satisfies the symmetry (1.2.39) and the f-sum rule (1.2.43).

Carrying out the q-sums in (2.1.30) within the elastic continuum model one finds for $T \ll T_D$ and $\mu_1 = \mu_2 = \mu \neq \mu_0$

$$\mathrm{Im}\,\Pi_2(q,\omega + i\delta) \propto$$

$$\omega \int_0^\infty d\omega' \, \omega'^4 \left(-\frac{\partial n^0(\omega')}{\partial \omega'}\right) \left\{ \arctan\left(\frac{c_\mu + c_{\mu_0}}{2c_{\mu_0}} \omega \tau_\mu(\omega')\right) \right.$$

$$\left. + \arctan\left(\frac{c_\mu - c_{\mu_0}}{2c_{\mu_0}} \omega \tau_\mu(\omega')\right) \right\} \qquad (2.1.32)$$

where we have introduced the lifetime of the mode

$$\tau_\mu(\omega) \equiv \left(\Gamma^{-1}_{\vec{q},\mu}\right)_{\omega_{\vec{q},\mu} = \omega} \qquad (2.1.33)$$

Since $\omega^4\left(-\frac{\partial n^0(\omega)}{\partial \omega}\right)$ is peaked at $\omega \sim 3T$ the integral can be approximated in replacing the bracket $\{\ \}$ by its value at the peak. This bracket then gives a correction factor to the Landau-Rumer behaviour $\alpha \propto \omega T^4$. In the collision-less limit, $\omega \tau_\mu(T) \gg 1$, $\{\ \} = \pi$ and the perturbation theoretic result is recovered. In the hydrodynamic limit, $\omega \tau_\mu(T) \ll 1$, one finds the result obtained from the Boltzmann-Peierls equation. This is remarkable since (2.1.30) took into account only the connections to the propagator lines according to fig. 4(a) whereas the 3-vertex corrections Γ_3 contained in fig. 4(b) were still left out. Now it is known[20] that the Boltzmann-Peierls equation is obtained by the ladder approximation of Γ_3, fig. 4(c) and leads both to first and second sound.

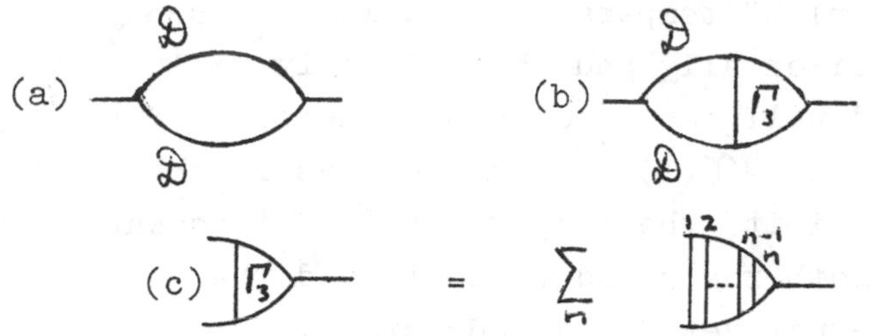

Figure 4.

The conclusion therefore is that while the result of the Boltzmann-Peierls equation is recovered for first sound without the inclusion of vertex corrections the second sound mode is completely missing in this approximation.

Finally we remark that in the case $\mu_0 = \mu = \ell$ the collinearity of the approximation (2.1.27) may be relaxed by inclusion of some dispersion,

$$\omega_{\vec{q}\ell} = c_\ell \, |\vec{q}| \, (1 - \gamma |\vec{q}|^2) \qquad (2.1.34)$$

Then the argument of the second arctan in the bracket $\{ \}$ of (2.1.30) is proportional to $(-\gamma)$ and this term still contributes. This case is of importance in superfluid Helium.

2. SOUND PROPAGATION IN SUPERFLUID HELIUM

2.1 Thermodynamic Relations[22)]

As in the case of the dielectric crystal in the distinction of the collision-less limit, $\omega\tau(T) \gg 1$, and the hydrodynamic limit, $\omega\tau(T) \ll 1$ ($\tau(T)$ is the relaxation time) is most important. In the latter limit superfluid Helium is well described by the two-fluid model which leads to first and second sound[6,23)]. This case, which is essentially a "high" temperature and low frequency limit, is experimentally and theoretically well explored[23)] and will not be discussed here. In the transition region $\omega\tau(T) \sim 1$, as well as in the collision-less limit, the temperature and frequency dependence of both the sound velocity and the attenuation are not yet fully understood.

Here we will limit ourselves to temperatures $T \lesssim 0.4^{\circ}K$ where the rotons are negligible so that the system is described by the phonon hamiltonian (1.1.46-53). In particular in the Nernst limit, $T \rightarrow 0$, useful information follows from thermodynamics.

Let C_v and C_p be the specific heats and S the entropy, all per unit mass, then

$$r \equiv \frac{C_p}{C_v} - 1 = \frac{T\alpha^2}{\rho C_v \kappa_T} = \frac{T\alpha^2 c^2}{C_p} \qquad (2.2.1)$$

where ρ is the mass density,

$$\alpha = -\frac{1}{\rho}\left(\frac{\partial \rho}{\partial T}\right)_p = -\rho\left(\frac{\partial S}{\partial p}\right)_T \qquad (2.2.2)$$

the expansion coefficient,

$$\kappa_{T,S} = \frac{1}{\rho}\left(\frac{\partial \rho}{\partial p}\right)_{T,S} \qquad (2.2.3)$$

the compressibilities and

$$c = (\rho \kappa_S)^{-\frac{1}{2}} \qquad (2.2.4)$$

the sound velocity. Use has been made of the relations

$$\frac{\kappa_T}{\kappa_S} = \frac{C_p}{C_V} \qquad (2.2.5)$$

In the Nernst limit $(T \rightarrow 0)$ we may write

$$C_V = C_p = \sigma(p)\, T^3 \tag{2.2.6}$$

so that from (2.2.1, 2) and with [23]

$$\begin{aligned} \rho_{T\to 0} &= 0.1450 \text{ g cm}^{-3} \\ c_{T\to 0} &= 2.383 \times 10^4 \text{ cm/sec} \\ \sigma &= 2.07 \times 10^5 \text{ cm}^2 \text{ sec}^{-2} \text{ K}^{-4} \\ (\alpha/T^3)_{T\to 0} &= 1.09 \times 10^{-3} \text{ K}^{-4} \end{aligned} \tag{2.2.7}$$

$$\left(\frac{r}{T^4}\right)_{T\to 0} = \frac{c^2 \rho^2 \alpha'^2(p)}{9\sigma(p)} = 3.25 \times 10^{-3} \text{ K}^{-4} \tag{2.2.8}$$

This result implies that, in the Nernst limit, it is unimportant whether the sound velocity (2.2.4) is defined adiabatically or isothermally. In the same limit we have from (2.2.3,4,8) and with $V = 1/\rho$

$$\left(\frac{\partial^2 S}{\partial V^2}\right)_T = \frac{\partial^2 p}{\partial T \partial V} = -2\rho^2 c \left(\frac{\partial c}{\partial T}\right)_\rho .$$

Because of (2.2.6) this has the consequence that

$$c \equiv c(T) - c(0) \propto T^4 \tag{2.2.9}$$

This result is of interest both for the analysis of data, (which would be more effective in the form $T^{-4}\Delta c(T)$ versus T instead of $\Delta c(T)$ versus T), and as a condition on the theory. In so far as this thermodynamic result is applicable to the collisionless limit (which is as questionable as the application of Landau's model of quantum hydrodynamics (1.1.46-53) in this limit). It excludes, for example, the theories of Hohenberg and Martin[24], who proposed $c(T) \propto \rho_S^{\frac{1}{2}}$ where ρ_S is the superfluid density, and of Etters[25] who found $c(T) \propto T^{5/2}$.

Another useful relation is obtained by making use of the formula of statistical mechanics for the specific heat of an ideal phonon gas, for which (1.1.50) holds,

$$C_V = \frac{1}{2\pi^2 \rho T^2} \int_0^\infty dq.q^3.\omega_q \frac{\partial \omega_q}{\partial q} \frac{e^{\omega_q/T}}{(e^{\omega_q/T} - 1)^2}$$

$$= \frac{2\pi^2}{45} \cdot \frac{T^3}{\rho c^3} \quad .$$

Comparison with (2.2.6) gives

$$\sigma(p) = \frac{2\pi^2}{45 \rho c^3} \quad . \tag{2.2.10}$$

This yields for the Grüneisen constant (1.1.43) in the Nernst limit

$$1 + 3u = - \left(\frac{\partial \log \sigma}{\partial \log \rho}\right)_{T \to 0} = - \frac{\rho c^2 \sigma'(p)}{\sigma(p)} \tag{2.2.11}$$

where we made use of (2.2.3,4). With (2.2.8) we find

$$(1 + 3u)^2 = \left(\frac{9c^2 r}{\sigma T^4}\right)_{T \to 0} = \left(\frac{9c^2(C_p - C_V)}{TC_V}\right)_{T \to 0} \tag{2.1.12}$$

which leads precisely to the Khalatnikov value[26]

$$u = 2.65 \quad . \tag{2.1.13}$$

The important feature of this result is that it was obtained in the Nernst limit while all other determinations of u were made at "high" temperatures and could therefore not be trusted at temperatures as low as $T \lesssim 0.4$ K.

In order to fix also the coupling parameter

(1.1.52) we have estimated[22)]

$$\rho_0 \frac{\partial u}{\partial \rho_0} \cong 3.0 \quad . \qquad (2.2.14)$$

2.2 Sound Velocity and Attenuation

The analogous calculation to (2.1.28-31) with the quantum hydrodynamic 3-phonon interaction (1.1.48-53) has been carried out by Pethick and ter Haar[27)] for a dispersion of the Khalatnikov form[26)] (2.1.34),

$$\omega_q = cq(1 - \gamma q^2). \qquad (2.2.15)$$

The cosine factor in $C^k_{-\vec{q},-\vec{q}',\vec{q}+\vec{q}'}$ is

$$(-\hat{q}' \cdot \frac{\vec{q}+\vec{q}'}{|\vec{q}+\vec{q}'|}) = -\frac{1 + (q/q')\zeta}{\sqrt{1 + 2\frac{q}{q'}\zeta + \frac{q^2}{q'^2}}} \qquad (2.2.16)$$

where $\zeta \equiv (\hat{q}.\hat{q}')$. If the collinearity condition discussed in Part II, section 1.2, is used, as was done by Pethick and ter Haar, then (2.2.16) takes the value ∓ 1 for $q' \gtrless q$. These authors get the value -3 instead, which leads to the coupling factor $(u + 1)^2$ in all their formulas. In order to understand this result let us formulate the problem more generally.

Introducing in addition to (1.3.8), (1.2.34) the Green functions and spectral functions

$$\mathcal{G}_{Q_q, -iP_{q'}}(\omega) \equiv \mathcal{D}'(q,\omega)\delta_{\vec{q},\vec{q}'} ; \; C^{\pm}_{Q_q,-iP_q}(\omega) \equiv B'^{\pm}(q,\omega)$$

$$\mathcal{G}_{iP^+_q,-iP_{q'}}(\omega) \equiv \mathcal{D}''(q,\omega)\delta_{\vec{q},\vec{q}'} ; \; C^{\pm}_{iP^+_q,-iP_q}(\omega) = B''^{\pm}(q,\omega)$$

(2.2.17)

we see that apart from vertex corrections and contributions from higher order vertices, $n \geqslant 4$ in (1.1.49), the following terms contribute to the self-energy Π_2 :

$$\Pi_2 = \Pi^{(3,3)} + 2\Pi^{(3,k)} + \Pi^{(k,k)} \qquad (2.2.18)$$

The corresponding diagrams are shown in fig. 5.

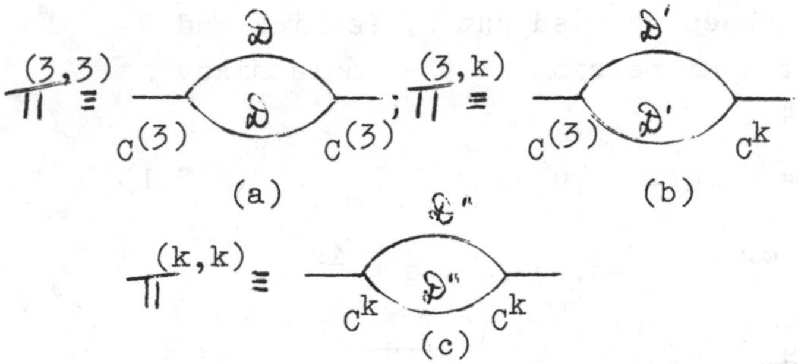

Figure 5.

It follows easily from (1.1.23), (1.2.27,28) that, in the approximation of phonon number conservation,

$$B''^{\pm}(q,\omega) = B^{\pm}(q,\omega) \qquad (2.2.19)$$

so that in this approximation $\Pi^{(3,3)}$ and $\Pi^{(k,k)}$ differ only through their coupling functions $C^{(3)}$ and C^{k}. The functions $B'^{\pm}(q,\omega)$, on the other hand, are different from $B^{\pm}$ and $B''^{\pm}$ since instead of (1.2.39,43) they satisfy

$$\left.\begin{aligned} B'^{-}(q,-\omega) &= +B'^{+}(q,\omega) \\ B'(q,-\omega) &= +B'(q,\omega) \end{aligned}\right\} \qquad (2.2.20)$$

and

$$\int_{-\infty}^{+\infty} B'(q,\omega)d\omega = \left\langle \left[iP_q, Q_q\right]_- \right\rangle = 1 \quad (2.2.21)$$

In order to estimate the importance of $\Pi^{(3,k)}$ we choose unperturbed functions $\mathcal{D}'^{0}$. This gives instead of (2.1.24)

$$L'^{0} \equiv 4\beta\sum_{\sigma'} \mathcal{D}'^{0}(q_1,-i\sigma')\mathcal{D}'^{0}(q_2,-i\sigma+i\sigma')$$

$$= \frac{n^0_{q_1} + n^0_{q_2} + 1}{i\sigma + \omega^0_{q_1} + \omega^0_{q_2}} + \frac{n^0_{q_1} + n^0_{q_2} + 1}{-i\sigma + \omega^0_{q_1} + \omega^0_{q_2}}$$

$$- \frac{n^0_{q_1} - n^0_{q_2}}{i\sigma - \omega^0_{q_1} + \omega^0_{q_2}} - \frac{n^0_{q_1} - n^0_{q_2}}{-i\sigma - \omega^0_{q_1} + \omega^0_{q_2}} . \quad (2.2.22)$$

Hence the first line of (2.1.24) and (2.2.22) which corresponds to <u>spontaneous</u> processes $(\rho_0 \rightleftharpoons \rho_1 + \rho_2)$ are identical while the second line, which corresponds to <u>induced processes</u> $(\rho_0 + \rho_1 \rightleftharpoons \rho_2)$ have opposite sign. We therefore introduce the following approximation for the full Lindhart function L':

$$L'_{spont} = L_{spont} \; ; \quad L'_{ind} = - L_{ind} . \quad (2.2.23)$$

Now the approximation of interest here is

$$cq \ll T \ll cq_0 \quad (2.2.24)$$

where q_0 is the cutoff wave number which is limited by the average interatomic distance

$$q_0 \lesssim \pi\left(\frac{\rho}{M}\right)^{1/3} = 0.877\ \text{Å}^{-1} . \quad (2.2.25)$$

The corresponding "Debye temperature" is

$$cq_0 \lesssim c\,\pi\left(\frac{\rho}{M}\right)^{1/3} = 16.0\ \text{K}. \quad (2.2.25')$$

With the condition (2.2.24) the spontaneous part $\Pi_{2,\text{spont.}}$ of (2.2.18) becomes essentially temperature independent, i.e., it only contributes to the sound velocity renormalisation at $T = 0$, [28]

$$c(0) - c^{0}(0) \equiv \left[\frac{\omega_q - \omega_q^{0}}{q}\right]_{q\to 0} = \Pi_{2,\text{spont.}} \tag{2.2.26}$$

This contribution is, however, difficult to calculate due to the angular dependence of $C^{(3)}$ and C^{k}.

Turning now to the discussion of the induced part $\Pi_{2,\text{ind.}}$ of (2.2.18) we have to come back to the collinearity condition. The latter is certainly justified for the attenuation as given by (2.1.19) since, under the condition (2.2.24), the term $(-\gamma q^2)$ in (2.2.15) should be negligible. It is, however, less obvious for the sound velocity renormalization, $\Delta c(T)$, as determined by the first line of (2.1.10). Indeed, while in equations (2.1.30, 31) approximate energy conservation results from the δ-function combined with the peaks of the spectral functions B, the δ-function is missing in Re Π. However, since the main contribution in the sum over $\vec{q}\,'$ comes from thermal phonons, $q' \cong 3T$, the condition (2.2.24) allows the development of (2.2.16)

$$(-\hat{q}\,' \cdot \frac{\vec{q} + \vec{q}\,'}{|\vec{q} + \vec{q}\,'|} = -1 + \frac{q^2}{q'^2}\zeta^2 + O(\frac{q^3}{q'^3}) \tag{2.2.27}$$

If, in the numerator of (2.1.28), all terms of order q/q' are neglected the collinearity condition is thus satisfied, and one finds in the approximation (2.2.19, 23)

$$\Pi_{2\ \text{ind.}} \mathrel{\hat{=}} (2u-1)^2\left[1 + \frac{6}{2u-1} + \frac{9}{(2u-1)^2}\right]\Pi^{(3,3)}_{\text{ind.}}$$

$$= 4(u-2)^2\,\Pi^{(3,3)}_{\text{ind.}} \quad . \qquad (2.2.28)$$

Here $(u+1)^2$ is the coupling factor of Pethick and ter Haar.

There remains the problem of higher order corrections, in particular vertex corrections. According to preliminary calculations of this author[22], vertex corrections seem to have an effect of the order of one on the value of the coupling factor. The most important vertex correction in the limit $\omega\tau \gg 1$ appeared to be the sum of bubble diagrams Σ shown in fig. 6.

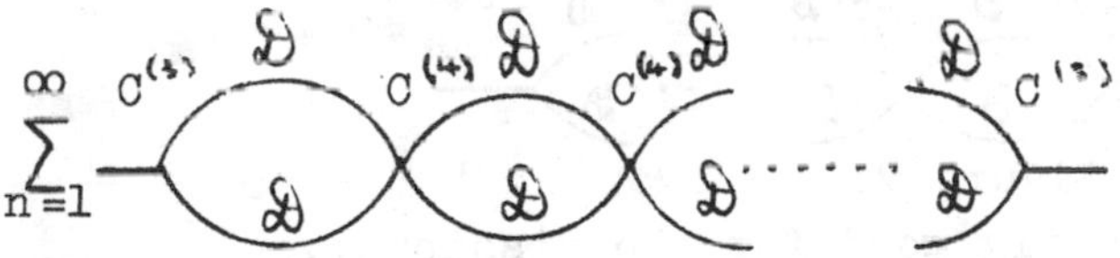

Figure 6.

where

$$\Sigma = \frac{\Pi^{(3,3)}}{1 + \Pi^{(3,3)}\dfrac{2C_4}{\beta C_3^2\ cq}} \qquad (2.2.29)$$

The reason for the belief that (2.2.29) gives the leading vertex correction in the limit $\omega\tau \gg 1$ is that these diagrams lead to a collective mode which has been studied in linear chain models[29]. It turns out that in superfluid Helium, at temperatures low enough so that $\omega\tau \gg 1$ is fulfilled,

($T \lesssim 0.4K$ for frequencies of the order of 100 Mc/sec) this collective mode cannot be excited but manifests itself through a substantial vertex correction.

This calculation is of course not very reliable, first because the bubble diagrams formed with $\mathcal{D}'$-lines at the extremities, see fig. 7 have been neglected. Second, the sound velocity renormalization at $T = 0$, (2.2.26), has not been treated properly because of the difficulties of calculation mentioned above. Third, ladder insertions into the bubbles, see fig. 8(a), have been excluded on the basis that ladders are typical hydrodynamic contributions leading to a Boltzmann equation and are negligible in the limit $\omega\tau \gg 1$ [20]. While this

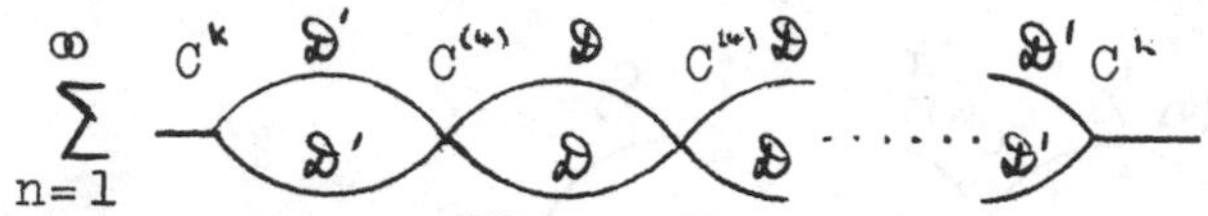

Figure 7.

argument is certainly good for the dissipative (i.e. imaginary) part of the self-energy, it is less obvious for the renormalization, i.e. real, part. In addition higher order diagrams such as that shown in fig. 8(b) may be important, too.

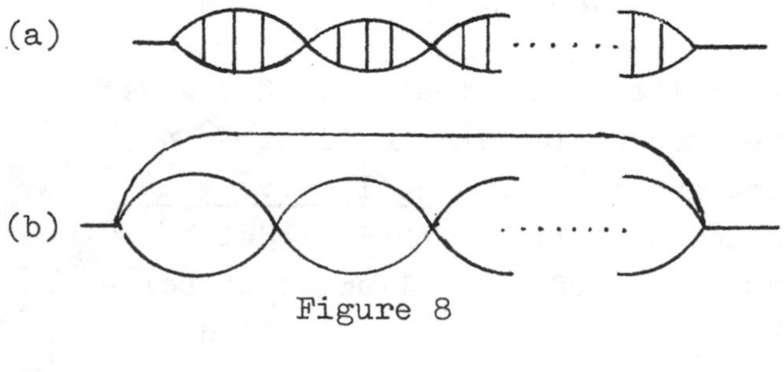

Figure 8

3. BOLTZMANN-PEIERLS EQUATION[30)]

3.1 Equations of Motion for Phonon Green Functions

From the phonon hamiltonian (1.1.17,18,19) we find in Heisenberg representation, with

$$H = H_{harm} + H_{anh} , \qquad (2.3.1)$$

$$\frac{\partial u_x(t)}{\partial t} = i\left[H, u_x\right]_- (t) = \frac{P_x(t)}{M_x} \qquad (2.3.2)$$

$$\frac{\partial P_x(t)}{\partial t} = i\left[H, P_x\right]_- (t) =$$

$$= \sum_{n \geqslant 2} \sum_{x_2 \ldots x_n} U_{x x_2 \ldots x_n} u_{x_2}(t) \ldots u_{x_n}(t) \qquad (2.3.3)$$

In the notation of Part I, section 4.2, we write $a(1) \equiv a_{x_1}(t_1)$ where a_x is u_x or P_x and $M_1 \equiv M_x$. The external interaction hamiltonian (1.3.22) may be written, (adding a factor $\sqrt{M_1}$ for convenience),

$$H'_t = \int_C d1 \sqrt{M_1}\, \phi(1) u(1)\, \delta(t_1 - t) , \qquad (2.3.4)$$

Then we have with the definitions (1.4.11, 14, 15)

$$\langle a_\phi (1)\rangle_\phi = \langle T_C(\mathbb{S}_C a(1))\rangle \,/\langle \mathbb{S}_C\rangle$$
$$= \langle \mathbb{S}_C(-\infty_-, t_1)\, a(1)\, \mathbb{S}_C(t_1, -\infty_+)\rangle \,/\langle \mathbb{S}_C\rangle . \qquad (2.3.5)$$

Using the analogue of (1.2.44),

$$\frac{\partial}{\partial t} \mathbb{S}_C(t, t_o) = -i\, H'_t\, \mathbb{S}_C(t, t_o) \qquad (2.3.6)$$

and its hermitian conjugate

$$\frac{\partial}{\partial t} \mathbb{S}_C(t_0,t) = i\, \mathbb{S}_C(t_0,t) H'_t \qquad (2.3.6')$$

differentiation of (2.3.5) with respect to t, gives

$$\frac{\partial}{\partial t_1} \langle a_\phi(1) \rangle_\phi = \left\langle \mathbb{S}_C(-\infty_-, t_1)\, i \left[H'_{t_1}, a(1) \right]_- \mathbb{S}_C(t_1, -\infty_+) \right\rangle .$$

$$. \langle \mathbb{S}_C \rangle^{-1} + \left\langle \left(\frac{\partial a}{\partial t_1}\right)_\phi (1) \right\rangle_\phi . \qquad (2.3.7)$$

Thus

$$\frac{\partial}{\partial t_1} \langle u_\phi(1) \rangle_\phi = \frac{1}{M_1} \langle P_\phi(1) \rangle_\phi \qquad (2.3.8)$$

$$\frac{\partial}{\partial t_1} \langle P_\phi(1) \rangle_\phi =$$

$$- \sqrt{M_1}\, \phi(1) - \sum_{n \geq 2} \frac{1}{(n-1)!} \sum_{x_2 .. x_n} U_{x_1 x_2 .. x_n} .$$

$$. \left\langle u_{x_2 \phi}(t_1) \ldots u_{x_n \phi}(t_1) \right\rangle_\phi . \qquad (2.3.8)$$

With the definition

$$U_n(1, \ldots n) \equiv \sqrt{M_1 \ldots M_n}\; U_{x_1 \ldots x_n} .$$

$$. \delta(t_1 - t_2)\, \delta(t_2 - t_3) \ldots \delta(t_{n-1} - t_n) \qquad (2.3.9)$$

the combination of the two equations (2.3.8) yields

$$-\sqrt{M_1}\, \frac{\partial^2}{\partial t_1^2} \langle u_\phi(1) \rangle_\phi = \phi(1) + \sum_{n \geq 2} \frac{1}{(n-1)!}$$

$$. \int d2 \ldots \int dn\, \sqrt{M_2 \ldots M_n}\; U_n(1,,,n) \left\langle T_C(u_\phi(2) \ldots u_\phi(n)) \right\rangle \qquad (2.3.10)$$

Acting on this equation with the functional derivative $\delta / \delta \phi(1')$ we obtain with (1.4.14,18) and using the definition

$$\mathbb{D}_m(1 \ldots m) \equiv \sqrt{M_1 \ldots M_m}\; \mathbb{G}_m^{irr}(1 \ldots m) - \frac{\partial^2}{\partial t_1{}^2} \mathbb{D}_2(1,1')$$

$$- \int d2\; U_2(1,2)\, \mathbb{D}_2(2,1')$$

$$= \delta(1-1') + (-i)^{-\frac{1}{2}} \frac{1}{2!} \int d2 \int d3\;\; U_3(1,2,3) \Big\{ \mathbb{D}_3(2,3,1')$$

$$+ \;\mathbb{D}_1(3)\; \mathbb{D}_2(2,1') + \mathbb{D}_1(2)\;\; \mathbb{D}_2(3,1') \Big\}$$

$$+ \;\; (-i)^{-1} \frac{1}{3!} \int d2 \int d3 \int d4\;\; U_4(1,2,3,4)\; .$$

$$K(2,3,4,1') + \ldots . \qquad (2.3.11)$$

where

$$K(2,3,4,1') \;=\; \mathbb{D}_4(2,3,4,1')$$

$$+ \sum_P \Big\{ \mathbb{D}_1(2)\;\; \mathbb{D}_3(3,4,1') + \mathbb{D}_2(2,3)\, \mathbb{D}_2(4,1')$$

$$+ \mathbb{D}_1(2)\;\; \mathbb{D}_1(3)\;\;\; \mathbb{D}_2(4,1') \Big\}$$

and Σ_P represents a sum over the cyclic permutations of the labels 2, 3, 4.

Defining the non-equilibrium m-point vertex function[31] as

$$\Gamma_m(1\ldots m) \equiv -\frac{\delta^m \mathbb{D}_2^{-1}(1,2)}{\delta\mathbb{D}_1(3)\ldots\delta\mathbb{D}_1(m)}$$

$$= \frac{\delta^{m-n}\Gamma_n(1\ldots n)}{\delta\mathbb{D}_1(n+1)\ldots\delta\mathbb{D}_1(m)} \qquad (2.3.13)$$

the higher order cumulants $\mathbb{D}_m$ factor into Γ-functions multiplied by $\mathbb{D}_2$- functions. Indeed, for $m = 3$, one finds

$$(-i)^{\frac{1}{2}}\mathbb{D}_3(1,2,3) = \frac{\delta\mathbb{D}_2(2,3)}{\delta\phi(1)} = \int dl' \frac{\delta\mathbb{D}_2(2,3)}{\delta\mathbb{D}(1')}\cdot\frac{\delta\mathbb{D}_1(1')}{\delta\phi(1)}$$

$$= (-i)^{\frac{1}{2}}\int dl' \frac{\delta\mathbb{D}_2(2,3)}{\delta\mathbb{D}_1(1')}\mathbb{D}_2(1,1') .$$

Combining this relation with the identity

$$\int dk'\left[\frac{\delta\mathbb{D}_2(k,k')}{\delta\mathbb{D}_1(m)}\mathbb{D}_2^{-1}(k',\ell) + \mathbb{D}_2(k,k')\frac{\mathbb{D}_2^{-1}(k',\ell)}{\delta\mathbb{D}_1(m)}\right] = 0$$

one finds

$$\mathbb{D}_3(1,2,3) =$$

$$\int dl'd2'd3'\mathbb{D}_2(1,1')\mathbb{D}_2(2,2')\mathbb{D}_2(3,3')\Gamma_3(1',2',3') \qquad (2.3.14)$$

This equation is shown in graphical form below.

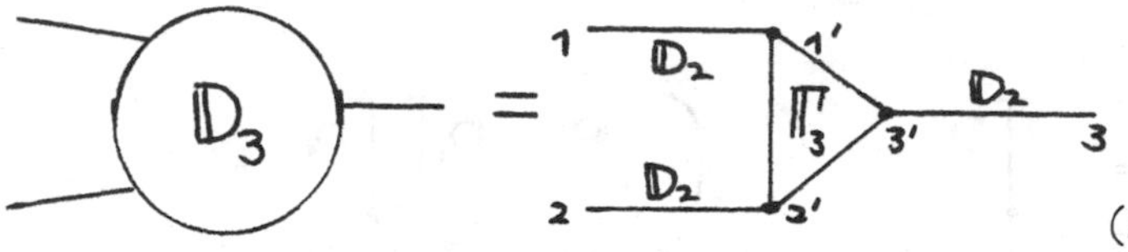

(2.3.14')

To lowest order in H_{anh}, Γ_m reduces to $U_m(1...m)$. This factorisation of $\mathbb{D}_2$-functions allows one to write (2.3.12) in the form[17)]

$$\int d2\ L(1,2)\,\mathbb{D}_2(2,1') = \delta(1-1') + \int d2\ \mathbb{M}(1,2)\,\mathbb{D}_2(2,1') \tag{2.3.15}$$

where

$$\begin{aligned} L(1,2) &\equiv -\delta(1-2)\frac{\partial^2}{\partial t_1^2} - U_2(1,2) \\ &= -\delta(t_1-t_2)\left\{\delta(x_1-x_2)\frac{\partial^2}{\partial t_1^2} + \sqrt{M_1M_2}\,U_{x_1x_2}\right\}. \end{aligned} \tag{2.3.16}$$

and $\mathbb{M}(1,2)$ is the <u>non-equilibrium self-energy</u>. Indeed, (2.3.15) is nothing else than the Dyson equation since, for $H_{anh} = 0$, one has $\mathbb{M}(1,2) = 0$ and (2.3.15) is then solved by

$$\mathbb{D}_2^{o}(1,2) = L^{-1}(1,2)$$

where the index o expresses, as previously, absence of the interaction H_{anh}. By comparison of (2.3.15) with (2.3.12) and using (2.3.14) one can obtain the self-energy to lowest order in H_{anh}[17)]. The result of such a calculation is expressed in the graphical equation below.

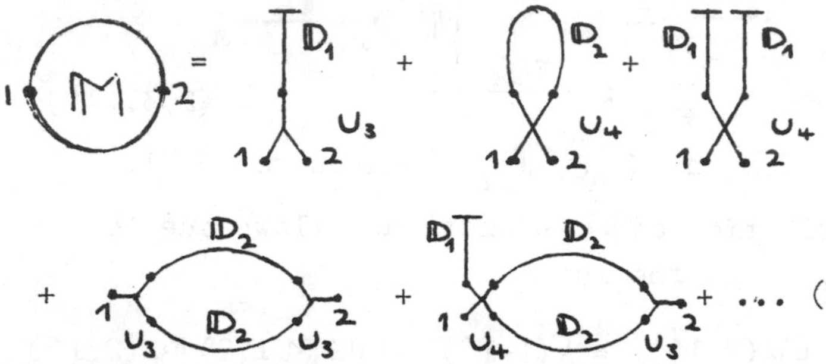

(2.3.18)

According to (1.4.25, 26) an equation of motion for the correlation function $\mathbb{D}^{<}(1,2) \equiv \sqrt{M_1 M_2}\, \mathbb{G}^{<}(1,2)$ may be obtained by acting on (2.3.10) with the functional derivative $\delta / \delta_{<}\phi(1')$. An easier way to get this equation is by using the definition analogous to (1.4.10)

$$\mathbb{D}_2(1,1') = \begin{cases} \mathbb{D}^{>}(1,1') \; ; & t_1 > t_{1'} \\ \mathbb{D}^{<}(1,1') \; ; & t_1 < t_{1'} \end{cases} \qquad (2.3.19)$$

and

$$\mathbb{M}(1,1') = \begin{cases} \mathbb{M}^{>}(1,1') \; ; & t_1 > t_{1'} \\ \mathbb{M}^{<}(1,1') \; ; & t_1 < t_{1'} \end{cases} \qquad (2.3.20)$$

(all time arguments being on the path C) and to split the integral over C in (2.3.15) into its ascending and descending parts. With $t_1 = t_{1+} > t_1' = t_{1'+}$ we get, remembering that according to (2.3.16) L(1,2) is local in time[17),]

$$\int d2\, L(1,2)\, \mathbb{D}^{>}(2,1') = \int_{-\infty}^{t_1'} d2\, \mathbb{M}^{>}(1,2)\, \mathbb{D}^{<}(2,1)$$
$$+ \int_{t_{1'}}^{t_1} d2\, \mathbb{M}^{<}(1,2)\, \mathbb{D}^{>}(2,1') + \int_{t_1}^{+\infty} d2\, \mathbb{M}^{<}(1,2)\, \mathbb{D}^{>}(2,1')$$

$$- \int_{-\infty}^{+\infty} d2\, \mathbb{M}^{<}(1,2)\, \mathbb{D}^{>}(2,1')$$

$$= \int_{-\infty}^{+\infty} d2 \left\{ \mathbb{M}^{>}(1,2)\, \mathbb{D}_2(2,1') - (\mathbb{M}(1,2) - \mathbb{M}^{>}(1,2)) \cdot \mathbb{D}^{>}(2,1') \right\} \qquad (2.3.21)$$

Using definitions analogous to (1.2.12),

$$\left.\begin{aligned} \mathbb{D}^{\text{ret}} &\equiv \mathbb{D}_2 - \mathbb{D}^{<} ; \mathbb{M}^{\text{ret}} \equiv \mathbb{M} - \mathbb{M}^{<} \\ \mathbb{D}^{\text{adv}} &\equiv \mathbb{D}_2 - \mathbb{D}^{>} ; \mathbb{M}^{\text{adv}} \equiv \mathbb{M} - \mathbb{M}^{>} \end{aligned}\right\} (2.3.22)$$

(2.3.21) may be written

$$\int d2\, L(1,2)\, \mathbb{D}^{>}(2,1') = \int_{-\infty}^{+\infty} d2 \left\{ \mathbb{M}^{\text{adv}}(1,2)\, \mathbb{D}^{>}(2,1') + \mathbb{M}^{>}(1,2)\, \mathbb{D}_2(2,1') - \mathbb{M}^{<}(1,2)\, \mathbb{D}^{>}(2,1') \right\} \qquad (2.3.23)$$

If in (2.3.15) we take $t_1 = t_{1+} < t_1 = t_{1'+}$ we obtain the analogous equation to (2.3.23)

$$\int d2\, L(1,2)\, \mathbb{D}^{<}(2,1') = \int_{-\infty}^{+\infty} d2 \left\{ \mathbb{M}^{\text{ret}}(1,2)\, \mathbb{D}^{<}(2,1') + \mathbb{M}^{<}(1,2)\, \mathbb{D}(2,1') - \mathbb{M}^{<}(1,2)\, \mathbb{D}^{>}(2,1') \right\} . \qquad (2.3.24)$$

Subtracting (2.3.24,2) from (2.3.15) we arrive at[17)]

$$\int d2\, L(1,2)\, \mathbb{D}^{P}(2,1') = \delta(1 - 1') + \int d2\, \mathbb{M}^{P}(1,2)\, \mathbb{D}^{P}(2,1') \qquad (2.3.25)$$

where P denotes either retarded or advanced.

3.2 Linearized Boltzmann-Peierls Equations

Making use of the symmetry

$$\begin{aligned} \mathbb{D}_2(1,1') &= \mathbb{D}_2(1',1) \\ \mathbb{M}(1,1') &= \mathbb{M}(1',1) \\ U_2(1,1') &= U_2(1',1) = U_2(1-1') \end{aligned} \tag{2.3.26}$$

and of (2.3.16), equation (2.3.15) may be written in the two alternative forms

$$-\frac{\partial^2}{\partial t_1^2}\mathbb{D}_2(1,1') = \delta(1-1') + \int d2\,\left[U_2(1-2)+\mathbb{M}(1,2)\right]\cdot \mathbb{D}_2(2,1')$$

$$-\frac{\partial^2}{\partial t_1^2}\mathbb{D}_2(1,1') = \delta(1-1') + \int d2\,\mathbb{D}_2(1,2)\cdot\left[U_2(2-1')+\mathbb{M}(2,1')\right].$$

Subtraction of these two equations and splitting the path C in its two parts leads to a generalized Boltzmann-Peierls equation

$$\begin{aligned} -\left(\frac{\partial^2}{\partial t_1^2} - \frac{\partial^2}{\partial t_{1'}^2}\right)\mathbb{D}_2(1,1') = {} & \int_{-\infty}^{+\infty} d2\left\{\left[U_2(1-2)+\mathbb{M}(1,2)\right]\mathbb{D}_2(2,1') - \right. \\ & \left. - \mathbb{D}_2(1,2)\left[U_2(2-1')+\mathbb{M}(2,1')\right]\right\} - \\ & - \int_{-\infty}^{+\infty} d2\left\{\mathbb{M}^<(1,2)\,\mathbb{D}^>(2,1') - \mathbb{D}^<(1,2)\,\mathbb{M}(2,1')\right. \end{aligned} \tag{2.3.27}$$

Now we make the assumption of linear response theory, namely that the external perturbation $\phi(1)$ is so weak that higher than linear terms can be dropped.

We write

$$\mathbb{D}_2(1, 1') = D_2(1 - 1') + \delta \mathbb{D}_2(1, 1')$$
$$\mathbb{M}(1, 1') = M(1 - 1') + \delta \mathbb{M}(1, 1') \qquad (2.3.28)$$

etc., where $D_2(1)$, $M(1)$ are the equilibrium functions which are homogeneous in $\vec{R}_I$ and t matrices in b, i, remembering that $x \equiv (I, b, i)$. Then (2.3.27) splits into the zeroth and first order equations

$$-\left(\frac{\partial^2}{\partial t_1{}^2} - \frac{\partial^2}{\partial t_1{}^2}\right) D(1-1') =$$
$$\int_{-\infty}^{+\infty} d2 \Big\{ \left[U_2(1-2) + M(1-2)\right] D_2(2-1')$$
$$- D_2(1-2) \left[U_2(2-1') + M(2-1')\right]\Big\} \qquad (2.3.29)$$
$$- \int_{-\infty}^{+\infty} d2 \left\{ M^<(1-2)\, D^>(2-1') - D^<(1-2) M^>(2-1')\right\}$$

which is the equilibrium equation of motion, and

$$-\left(\frac{\partial^2}{\partial t_1{}^2} - \frac{\partial^2}{\partial t_1{}^2}\right) \delta \mathbb{D}_2(1, 1')$$
$$= \int_{-\infty}^{+\infty} d2 \Big\{ \delta\mathbb{M}(1,2) D_2(2-1') + \left[U_2(1-2) + M(1-2)\right] \delta \mathbb{D}_2(2,1')$$
$$- D_2(1-2) \delta\mathbb{M}(2,1') - \delta\mathbb{D}_2(1,2)\left[U_2(2-1') + M(2-1')\right]\Big\}$$
$$- \int_{-\infty}^{+\infty} d2 \Big\{ \delta\mathbb{M}^<(1,2) D^>(2-1') + M^<(1-2)\, \delta\mathbb{D}^>(2,1')$$
$$- D^<(1-2)\, \delta\mathbb{M}^>(2,1') - \delta\mathbb{D}^<(1,2) M^>(2-1')\Big\} \qquad (2.3.30)$$

which is a generalized linearized Boltzmann-Peierls equation. We now make the Fourier transformation in space (1.1.20, 20′) and in time using the following notation

$$u(1) = \int d\tilde{1}\, f(1,\tilde{1})u(\tilde{1}) \qquad (2.3.31)$$

where

$$u(\tilde{1}) = \int dt\, Q_{q_{\tilde{1}}}(t) e^{i\omega_{q_{\tilde{1}}} t} \qquad (2.3.31')$$

$$f(1,\tilde{1}) = \frac{1}{2\pi}\left(NM_{b_1}\omega^{o}_{q_{\tilde{1}}}\right)^{-\frac{1}{2}} e_{b_1 i_1}(q_{\tilde{1}}) e^{i(\vec{q}_{\tilde{1}}\cdot\vec{R}_{l_1} - \omega_{\tilde{1}} t_1)} \qquad (2.3.31'')$$

and $\int d\tilde{1} \equiv \int d\omega_{\tilde{1}} \sum_{q_{\tilde{1}}}$. The quantities with two arguments then transform as

$$A(1,2) = \int d\tilde{1} \int d\tilde{2}\, f(1,\tilde{1}) f(2,\tilde{2}) A(\tilde{1}, \tilde{2}) \qquad (2.3.32)$$

A function which is homogeneous in space-time has a diagonal Fourier transform which, however, is still a matrix in the polarization and mode index ,

$$A(1-2) \longrightarrow A(\tilde{1})\, \delta(\tilde{1}, \tilde{2}) \qquad (2.3.33)$$

In particular, according to (1.1.21, 21′)

$$U_2(1-2) \rightarrow \omega^{o2}(\tilde{1})\, \delta(\tilde{1}, \tilde{2}) \qquad (2.3.33')$$

where $\omega^{o}(\tilde{1}) \equiv \omega^{o}_{q_{\tilde{1}}}$.

In this notation the Fourier transform of (2.3.30) is

$$\left\{\omega^2_{\tilde{1}} - \omega^2_{\tilde{1}'} - (\omega^{o2}(\tilde{1}) + M(\tilde{1})) + (\omega^{o2}(\tilde{1}') + M(\tilde{1}')\right\}\delta D_2(\tilde{1},\tilde{1}')$$

$$= \left[D_2(\tilde{1}') - D_2(\tilde{1})\right]\delta M(\tilde{1},\tilde{1}')$$

$$+\left[M^{>}(\tilde{1}')\,\delta\mathbb{D}^{<}(\tilde{1},\tilde{1}') - M^{<}(\tilde{1})\,\delta\mathbb{D}^{>}(\tilde{1},\tilde{1}')\right.$$
$$\left.+\ D^{<}(\tilde{1})\,\delta\mathbb{M}^{>}(\tilde{1},\tilde{1}') - D^{>}(\tilde{1}')\,\delta\mathbb{M}^{<}(\tilde{1},\tilde{1}')\right] \tag{2.3.34}$$

We consider only the part of (2.3.34) which is diagonal in μ and write

$$\omega_{\tilde{1}} = \omega + \frac{\nu}{2}\,; \qquad \omega_{\tilde{1}'} = \omega - \frac{\nu}{2}$$
$$q_{\tilde{1}} = (\vec{q} + \frac{\vec{p}}{2}, \mu); \qquad q_{\tilde{1}'} = (\vec{q} - \frac{\vec{p}}{2}, \mu) \tag{2.3.35}$$

where the external variables $\vec{p}, \nu$ are supposed to be small, corresponding to slowly varying space-time perturbations. In addition we introduce the quasi-particle renormalization (2.1.9),

$$\left.\begin{aligned} \omega^{o2}(\tilde{1}) + M(\tilde{1}) &= (\omega_{\vec{q}+\frac{\vec{p}}{2},\mu} + i\,\Gamma_{\vec{q}+\frac{\vec{p}}{2},\mu})^2 \\ \omega^{o2}(\tilde{1}') + M(\tilde{1}') &= (\omega_{\vec{q}-\frac{\vec{p}}{2},\mu} + i\,\Gamma_{\vec{q}-\frac{\vec{p}}{2},\mu})^2 \end{aligned}\right\} \tag{2.3.36}$$

where the widths Γ_q are also supposed to be small, so that the <u>conditions of the hydrodynamic limit</u> are fulfilled,

$$\Gamma_q \ll \omega_q\,; \quad \nu \ll \omega_q\,; \quad |\vec{v}_q \cdot \vec{p}| \ll \omega_q \tag{2.3.37}$$

where $\vec{v}_q$ is the group velocity. Then the first bracket in (2.3.34) becomes to first order in the small quantities

$$\{\ \} = 2\omega\nu - 2\,\omega_q(\vec{v}_q \cdot \vec{p}) \tag{2.3.38}$$

which is recognized as the <u>kinematical factor</u> of the Boltzmann-Peierls equation.

The last bracket, [ ] , in (2.3.34) leads to the collision integral of the Boltzmann-Peierls equation. To see this we use (2.3.18, 20) to write $\mathbb{M}^{\gtrless}$ graphically to lowest order in H_{anh}

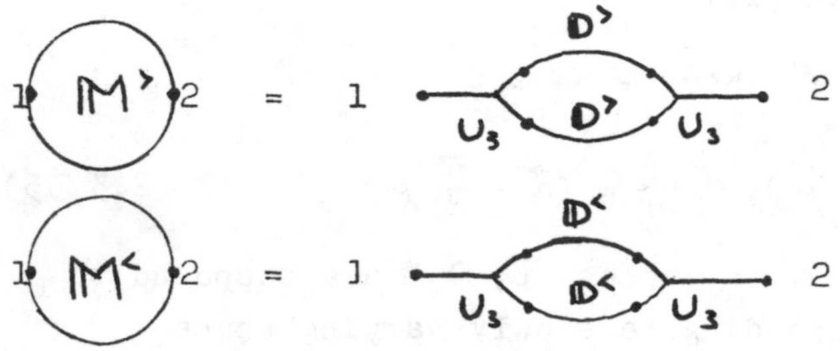

(2.3.39)

In the linear response approximation and in Fourier transform this leads to

$$M^{\gtrless}(\tilde{1}) = \int d\tilde{2} \int d\tilde{3} \; D^{\gtrless}(\tilde{2})\, D^{\gtrless}(\tilde{3}) . \left| U_3(\tilde{1}, \tilde{2}, \tilde{3})\right|^2$$

and

(2.3.40)

$$\delta \mathbb{M}^{\gtrless}(\tilde{1},\tilde{1}') =$$

$$\int d\tilde{2} \int d\tilde{3} \int d\tilde{3}' D^{\gtrless}(\tilde{2})\, \delta \mathbb{D}^{\gtrless}(\tilde{3},\tilde{3}')\; U_3(\tilde{1},\tilde{2},\tilde{3}) U_3(\tilde{1}',\tilde{2}',\tilde{3}')$$

$$+ \int d\tilde{2} \int d\tilde{2}' \int d\tilde{3}\, \delta \mathbb{D}(\tilde{2},\tilde{2}')\; D^{\gtrless}(\tilde{3}) U_3(\tilde{1},\tilde{2},\tilde{3}) U_3(\tilde{1}',\tilde{2}',\tilde{3}') .$$

(2.3.41)

Writing as in (2.3.35)

$$\omega_{\tilde{2}} = \omega' + \frac{\nu}{2} \; ; \quad \omega_{\tilde{2}'} = \omega' - \frac{\nu}{2}$$

$$q_{\tilde{2}} = (\vec{q}' + \frac{\vec{p}}{2}, \mu') ; \; q_{\tilde{2}'} = (\vec{q}' - \frac{\vec{p}}{2}, \mu')$$

$$\omega_{\tilde{3}} = \omega'' + \frac{\nu}{2} \; ; \quad \omega_{\tilde{3}'} = \omega'' - \frac{\nu}{2}$$

$$q_{\tilde{3}} = (\vec{q}'' + \frac{\vec{p}}{2}, \mu'') \; ; \; q_{\tilde{3}'} = (\vec{q}'' - \frac{\vec{p}}{2}, \mu'') \quad ,$$

taking the limit $\vec{p} \to 0,\ \nu \to 0$ except in the factors $\delta \mathbb{D}^{\gtrless}$ and keeping only diagonal elements in (2.3.40, 41) the last bracket in (2.3.34) becomes

$$[\ ] = \sum_{q'} \sum_{q''} \int d\omega' \int d\omega'' \left| U_3(q\omega,\ q'\omega', q''\omega'') \right|^2 .$$

$$\cdot\ B(q\omega\,;\ q',\omega';\ \ q'',\omega''\,;\ p,\nu\ ) \qquad (2.3.42)$$

where

$$B(q,\omega;\ q',\omega';\ q'',\omega'';\ p,\nu)$$

$$= D^{>}(q'\omega')D^{>}(q''\omega'')\delta\mathbb{D}^{<}(q\omega;\ \vec{p}\nu\ )$$
$$-\ D^{<}(q'\omega')D^{<}(q''\omega'')\delta\mathbb{D}^{>}(q\omega;\ \vec{p}\nu\ )$$
$$+\ D^{<}(q\omega)\ \ D^{>}(q'\omega')\,\delta\mathbb{D}^{>}(q''\omega'';\ \vec{p}\nu)$$
$$-\ D^{>}(q\omega)\ \ D^{<}(q'\ \omega')\,\delta\mathbb{D}^{<}(q''\omega'';\ \vec{p}\nu\ )$$
$$+\ D^{<}(q\omega)\ \ D^{>}(q''\omega'')\,\delta\mathbb{D}^{>}(q'\ \ \omega'\ ;\ \vec{p}\nu)$$
$$-\ D^{>}(q\omega)\ \ D^{<}(q''\ \omega'')\,\delta\mathbb{D}^{<}(q'\omega'\,;\ \vec{p}\nu\ )$$

In order to obtain a transport equation a phonon distribution function has to be defined. This is achieved by generalizing the relations (1.2.27, 28, 34, 38) to the non-equilibrium situation writing

$$\mathbb{D}^{>}(\tilde{1},\tilde{1}') \ \equiv \mathbb{D}^{>}(q\omega;\ \vec{p}\nu\ ) = -2\pi i B(q\omega;\ \vec{p}\nu\ )(1+n(q\omega;\ \vec{p}\nu)) \qquad (2.3.43)$$

$$\mathbb{D}^{<}(\tilde{1},\tilde{1}') \ \equiv \mathbb{D}^{<}(q\omega;\ \vec{p}\nu\ ) = +2\pi i B(q\omega;\ \vec{p}\nu\ )n(q\omega;\ \vec{p}\nu)$$

This defines simultaneously the <u>non-equilibrium distribution function</u> <u>$n(q\omega\,;\ \vec{p}\nu)$</u> and <u>spectral function</u> <u>$B(q\omega\,;\ \vec{p}\nu\ )$</u>. In the linear response approximation

$$\left.\begin{aligned} n(q\omega;\vec{p}\nu) &= n^0(\omega) + \delta n(q\omega;\vec{p}\nu) \\ B(q\omega;\vec{p}\nu) &= B(q,\omega) + \delta B(q\omega;\vec{p}\nu) \end{aligned}\right\} \qquad (2.3.44)$$

Inserted into (2.3.43) this gives

$$\begin{aligned} \delta \mathbb{D}^{>}(q\omega;\vec{p}\nu) &= \\ &-2\pi i\left\{\delta B(q\omega;\vec{p}\nu)(1+n^0(\omega)) + B(q\omega)\delta n(q\omega;\vec{p}\nu)\right\} \\ \delta \mathbb{D}^{<}(q\omega;\vec{p}\nu) &= \\ &+2\pi i\left\{\delta B(q\omega;\vec{p}\nu)n^0(\omega) + B(q\omega)\delta n(q\omega;\vec{p}\nu)\right\} \end{aligned} \qquad (2.3.45)$$

Generalizing the dispersion relation (1.2.30) to the non-equilibrium situation (the Fourier transform of the time t of the total hamiltonian $H + H_t'$ is the external frequency ν and does not affect the analyticity in the internal frequency ω which is proper to the system).

We have

$$\delta \mathbb{D}_2(q\omega;\vec{p}\nu) = \frac{1}{2\pi i}\int_{-\infty}^{+\infty} d\omega'\left\{\frac{\delta \mathbb{D}^{>}(q\omega';\vec{p}\nu)}{\omega' - \omega - i\delta} - \frac{\delta \mathbb{D}^{<}(q\omega';\vec{p}\nu)}{\omega' - \omega + i\delta}\right\} \qquad (2.3.46)$$

and inserting (2.3.45)

$$\begin{aligned} \delta \mathbb{D}_2(q\omega;\vec{p}\nu) &= \\ &= -2\pi i\, B(q\omega)\delta n(q\omega;\vec{p}\nu) - \int_{-\infty}^{+\infty} d\omega'\,\delta B(q\omega';\vec{p}\nu)\,. \\ &\cdot\left\{\frac{1+n^0(\omega')}{\omega' - \omega - i\delta} - \frac{n^0(\omega')}{\omega' - \omega + i\delta}\right\} \end{aligned} \qquad (2.3.47)$$

Making use of (2.3.45) we write (2.3.42) as the sum of the collision term C and "inhomogeneous" term I,

$$\frac{1}{2\pi i}[\ ] = C[\delta n] + I[\delta B] \tag{2.3.48}$$

Inserting now (2.3.38, 47, 48) back into (2.3.34) and writing

$$D_2(\hat{1}') - D_2(\hat{1}) = \frac{\partial D_2(q\omega)}{\partial \vec{q}} \cdot \vec{p} + \frac{\partial D_2(q\omega)}{\partial \omega}\nu \tag{2.3.49}$$

we obtain after re-arrangement of terms and division by $2\pi i$

$$-2(\omega\nu - \omega_q(\vec{v}_q \cdot \vec{p}))B(q\omega)\delta n(q\omega; \vec{p}\nu) - C[\delta n]$$

$$= 2\left(\omega\nu - \omega_q(\vec{v}_q \cdot \vec{p})\right) \frac{1}{2\pi i}\int d\omega' \left\{\frac{1 + n^0(\omega')}{\omega' - \omega - i\delta} - \frac{n^0(\omega')}{\omega' - \omega + i\delta}\right\}$$

$$\cdot\ \delta B(q\omega'; \vec{p}\nu)$$

$$+ \ I[\delta B] + \frac{1}{2\pi i} \frac{\partial D_2(q\omega)}{\partial \vec{q}} \cdot \vec{p} + \frac{\partial D_2(q\ )}{\partial \omega}\nu \ .$$

$$\cdot\ \delta M(q\omega; \vec{p}\nu). \tag{2.3.50}$$

This is the generalized Boltzmann-Peierls equation. Its right hand side contains all the "inhomogeneous" terms, i.e. terms which are proportional to the external driving force $\phi(1)$. Indeed, to lowest order in H_{anh} the first order linear response term of (2.3.18) reads, if use is made of (1.4.19), (2.3.11),

$$\delta M(q\omega; \vec{p}\nu) = \int d2\ U_3(\hat{1}, \hat{1}', 2)\sqrt{M_2}\ (-i)^{\frac{1}{2}}\langle u_\phi(2)\rangle_\phi \tag{2.3.51}$$

On the other hand it follows from (2.3.45) that

$$-2\pi i\, \delta B(q\omega;\vec{p}\nu) = \delta \mathbb{D}^{>}(q\omega;\vec{p}\nu) - \delta \mathbb{D}^{<}(q\omega;\vec{p}\nu)$$
$$= \delta \mathbb{D}^{ret}(q\omega;\vec{p}\nu) - \delta \mathbb{D}^{adv}(q\omega;\vec{p}\nu) \tag{2.3.52}$$

But from the first order linear response terms of (2.3.25) one obtains by a subtraction analogous to that leading to (2.3.30)

$$-\left(\frac{\partial^2}{\partial t_1^2} - \frac{\partial^2}{\partial t_{1'}^2}\right)\delta \mathbb{D}^P(1,1')$$
$$= \int d2 \Big\{ \delta \mathbb{M}^P(1,2) D_2(2-1') + \left[U_2(1-2) + M^P(1-2)\right] \delta \mathbb{D}^P(2,1')$$
$$- D^P(1-2)\,\delta \mathbb{M}^P(2,1') - \delta \mathbb{D}^P(1,2)\left[U_2(2-1') + M^P(2-1')\right]\Big\} \tag{2.3.53}$$

Making use of the fact that in the approximation of (2.3.51)

$$\left.\begin{array}{l} \delta \mathbb{M}^{ret} - \delta \mathbb{M} = -\delta \mathbb{M}^{<} \cong 0 \\ \delta \mathbb{M}^{adv} - \delta \mathbb{M} = -\delta \mathbb{M}^{>} \cong 0 \end{array}\right\} \tag{2.3.54}$$

one sees by Fourier-transforming (2.3.53) that $\delta \mathbb{D}^P(q\omega;\vec{p}\nu)$ and through (2.3.52) also $\delta B(q\omega;\vec{p}\nu)$ are indeed expressible entirely in terms of (2.3.51).

Finally we remark that in order to obtain the usual Boltzmann-Peierls equation one has to use for the equilibrium spectral function $B(q\omega)$ in (2.3.50) its unperturbed form (1.2.41). The phonon distribution function (2.3.44) then takes the usual form

$$n(q, +\omega; \vec{p}\nu) = n_q^0 + \delta n_q(\vec{p}\nu)$$

$$n(q, -\omega; \vec{p}\nu) = -(1+n(q,+\omega; \vec{p}\nu)$$

$$-(1+n_q^0) - \delta n_q(\vec{p}\nu) \qquad (2.3.55)$$

Acknowledgment. I would like to thank Miss Thérese Scheuring for her comprehensive and efficient co-operation in typing the original manuscript.

REFERENCES

1) J.M. Ziman, "Electrons and Phonons", Oxford University Press (London 1960).

2) M. Born and K. Huang, "Dynamical Theory of Crystal Lattices", Clarendon Press (Oxford 1954).

3) C.P. Enz, "Electron-Phonon and Phonon-Phonon Interactions", in "Theory of Condensed Matter", International Atomic Energy Agency (Vienna 1968) pp. 729-782.

4) C.P. Enz, Ann. Phys. (N.Y.) 46, 114 (1968).

5) See ref. 2, § 14 and ref. 3, p. 736.

6) F. London, "Superfluids", Dover Publications (New York 1964), Vol. II, § 17.

7) A.A. Abrikosov, L.P. Gorkov and I.YE. Dzyaloshimskii, "Field Theoretical Methods in Statistical Physics" (ed. D. ter Haar), Pergamon Press (London 1965).

8) L.P. Kadanoff and G. Baym, "Quantum Statistical Mechanics", W.A. Benjamin, Inc. (New York 1962).

9) See, e.g. ref. 4, appendix A.

10) See, e.g. N.N. Bogoliubov and D.V. Shirkov, "Introduction to the Theory of Quantized Fields", Interscience Publishers, Ltd. (New York 1959).

11) A.I. Alekseev, Soviet Physics-Uspekhi 4, 23 (1961).

12) R. Kubo, J. Phys. Soc. Japan 12, 570 (1957).

13) M. Gaudin, Nucl. Phys. 15, 89 (1960).

14) C.P. Enz, Helv. Phys. Acta 38, 150 (1965).

15) L.V. Keldysh, Soviet Physics - JETP 20, 1018 (1965). P.C. Martin and J. Schwinger, Phys. Rev. 115, 1342 (1959). See also J. Schwinger, J. Math. Phys. 2, 407 (1961).

16) T. Högberg, Arkiv Fysik 29, 519 (1965); 34, 121 (1967). R.K. Wehner, Phys. Stat. Solidi 22, 527 (1967).

17) G. Niklasson and A. Sjölander, Ann. Phys. (N.Y.) 49, 249 (1968). See also B. Biezzerdes and D.F. DuBois; Phys. Rev. 168, 233 (1968); P. Wölfle, Z. Physik (to be published).

18) P.C.K. Kwok, "Green's Functions Method in Lattice Dynamics", Solid State Physics (ed. F. Seitz, D. Turnbull and H. Ehrenreich), Academic Press (New York and London 1967), Vol. 20, p. 213.

19) P.C. Kwok and P.C. Martin, Phys. Rev. 142, 495 (1966).

20) L.J. Sham, Phys. Rev. 156, 494 (1967); 163, 401 (1967).

21) R. Klein, Phys. Cond. Matter 6, 38 (1967).

22) C.P. Enz, Nuovo Cimento (in press).

23) J. Wilks, "The Properties of Liquid and Solid Helium", Clarendon Press (Oxford 1967).

24) P.C. Hohenberg and P.C. Martin, Phys. Rev. Letters 12, 69 (1964).

25) R.D. Etters, Phys. Rev. Letters 16, 119 (1966); Nuovo Cimento 44B, 68 (1966).

26) I.M. Khalatnikov, "Introduction to the Theory of Superfluidity", W.A. Benjamin Inc. (New York 1965).

27) J. Pethick and D. ter Haar, Physica 32, 1905 (1966). See also, J. Jäckle, Z. Physik (to be published).

28) S. Eckstein and B.B. Varga, Phys. Rev. Letters 21, 1311 (1968).

29) C.P. Enz, Bull. Am. Phys. Soc. 9, 278 (1964) and in "Mathematical Methods in Solid State and Superfluid Theory", Scottish Universities' Summer School, St. Andrews 1967 (ed. R.C. Clark and G.H. Derrick) Oliver & Boyd (Edinburgh 1969), p. 339; C.P. Enz and J.P. Müller, Phys. Cond. Matter (to be published).

30) Papers using equilibrium Green functions: L.J. Sham, ref. 20, (imaginary-time functions), R. Klein and R.K. Wehner, Phys. Cond. Matter 8, 141 (1968) and preprint (real-time functions). Papers using non-equilibrium Green functions: C. Horie and J.A. Krumhansl, Phys. Rev. 136, A 1397 (1964); P.C. Kwok and P.C. Martin, ref. 19; W. Götze and K.H. Michel, Phys. Rev. 157, 738 (1967) and preprint; P. Ieier, Phys. Cond. Matter 8, 241 (1968); G. Niklasson and A. Sjölander, ref. 17 (double real-time functions).

31) See, e.g. R.K. Wehner, ref. 16.

KINEMATICAL PROPERTIES OF EQUILIBRIUM STATES

E.J. Verboven

Instituut voor Theoretische Fysika
Universiteit Nijmegen

1. INTRODUCTION

The aim of Statistical Mechanics is to understand the macroscopic behaviour of matter from its microscopic structure. Macroscopic matter in equilibrium can occur in different phases: the gaseous phase, the liquid phase and the solid phase.

The usual approach to statistical mechanics is sufficient for the understanding of the properties of individual phases, as witnessed by the fairly well developed theories of gases and liquids in equilibrium, and by solid state physics.

An important problem, as yet not solved in the usual approach of statistical mechanics, is however the following. Let us consider a large system with two-body forces, depending only upon the distance between particles. How can one understand that such a system can occur in different phases, such as the gaseous, liquid and solid phase? The usual approach of statistical mechanics has not yet been able to give a satisfactory answer to this question. Ordinary ensemble theory predicts expectation values which are Euclidean invariant even when the temperature and chemical potential are such that the system should be crystalline. These remarks suffice to indicate that some of the basic problems in Statistical Mechanics are as yet unsolved. In an attempt to bring these problems to a solution new methods have been developed. One of these methods, the algebraic approach, and more specifically the approach in terms of C^*-algebras and their representations, will be described in these notes in order to discuss properties of matter in thermal equilibrium.

The analysis will be given for infinite quantum systems. Every real system is of course finite, but as long as one is concerned with bulk properties which are volume independent one can consider the system as well infinitely large. There are several fundamental theoretical reasons to do this. First of all it turns out that the notion of ergodicity only makes sense for a quantum system when the system is infinitely large, since only in such a situation can one make a consistent partition into local and global observables. Secondly, the Euclidean symmetry is present from the beginning, without the need to have recourse to boundary conditions. Finally, one has only sharply separated phases when a system is infinitely large.

Recently new methods have been developed in order to deal with infinite systems. Mathematically the need for these methods originates from the fact that the expectation values of observables for an infinite system can no longer be calculated from density matrices on Fock space. In the algebraic approach one considers therefore no longer the set of observables as a concrete algebra on Fock space but rather as an abstract C^*-algebra with a whole set of different representations suitable for the different densities and temperatures, including $T = 0$

i.e. the ground state. (The Fock representation belongs then to zero density.) In a representation, corresponding to a particular density and temperature, the expectation value (at this density and temperature) of an observable takes again a simple form. A second important advantage of the algebraic approach is, that it provides us with a mathematical frame which is very well suited in order to translate the classical ergodic theory into the quantum case.

2. MATHEMATICAL PRELIMINARIES

2.1 Definition of a C^*-algebra

In this chapter we expose briefly the mathematical methods needed in order to formulate the algebraic approach of statistical mechanics.

An <u>algebra</u> A is a set of elements satisfying the following properties:

(a) A is a vector space, i.e. for $a, b \in A$ and λ a complex number, $a + b \in A$ and $\lambda a \in A$.

(b) A multiplication operation is defined in A such that, for $a, b, c \in A$ and λ a complex number, $ab \in A$, and

$\lambda(ab) = (\lambda a)b = a(\lambda b)$ (homogeneity)

$a(b + c) = ab + ac$

$(a + b)c = ac + bc$ (distributivity)

and finally the associative law $a(bc) = (ab)c$.

If $ab = ba$ for all $a, b \in A$, the algebra is said to be <u>abelian</u>. Two elements $a, b \in A$ are said to commute if $ab = ba$.

The <u>centre</u> Z of an algebra A is the set of elements belonging to A, which commute with all the elements of A, i.e.

$$Z = \{c \mid c \in A, \quad ca = ac \text{ for all } a \in A\}.$$

Z is an abelian subalgebra of A.

An element $e \in A$ is called an identity of A if $ae = ea = a$ for all $a \in A$. This element is unique, because if e' would be another identity then $e'e = ee' = e = e'$.

A subset J of an algebra A is called a <u>left ideal</u> (resp. <u>right ideal</u>) if

(1) J is a subspace of A.

(2) $x \in J$ and $a \in A$ imply $ax \in J$ (resp. $xa \in J$). One writes $AJ \subset J$ (resp. $JA \subset J$).

If J is both a left and right ideal, it is called a <u>two-sided ideal</u>. Every algebra contains the trivial ideals $\{A\}$ and $\{0\}$. Every ideal different from $\{A\}$ and $\{0\}$ is called proper. An algebra is called <u>simple</u> if it contains no two-sided ideals different from $\{0\}$.

An element $i \in A$ is called <u>idempotent</u> if $ii = i$.

A set A of elements is called an <u>algebra with involution</u> or <u>*-algebra</u> if

(1) A is an algebra.

(2) An operation is defined in A, which assigns to each element a in A an element a^* in A in such a way that the following conditions are satisfied:

(a) $(\lambda a + \mu b)^* = \bar{\lambda} a^* + \bar{\mu} b^*$ where $\bar{\lambda}, \bar{\mu}$ are the complex conjugates of λ, μ resp.

(b) $a^{**} = a$.

(c) $(ab)^* = b^* a^*$.

a^* is called the <u>adjoint</u> of a.

An element a is called <u>self-adjoint</u> or <u>hermitian</u> if $a^* = a$.

An element a is called <u>normal</u> if $a^* a = aa^*$. A subset $S \in A$ is called self-adjoint if $S^* = S$, where $S^* = \{ a^* \mid a \in S\}$. Every self-adjoint ideal is a two sided ideal.

Every element a of a $*$-algebra A can be represented uniquely in the form

$$a = a_1 + i\, a_2$$

where a_1 and a_2 are hermitian elements of A. Indeed, take

$$a_1 = \frac{a + a^*}{2}, \qquad a_2 = \frac{a - a^*}{2i}.$$

A mapping β of an algebra A_1 into an algebra A_2 is called a <u>homomorphism</u> if

(1) $\beta[\lambda a] = \lambda \beta[a]$ for all $a \in A_1$

(2) $\beta[a + b] = \beta[a] + \beta[b]$ for all $a, b \in A_1$

(3) $\beta[a.b] = \beta[a] . \beta[b]$. for all $a, b \in A_1$.

It is a <u>$*$-homomorphism</u> if A_1 and A_2 are $*$-algebras and if

(4) $\beta[a^*] = \beta[a]^*$.

The inverse image of the zero of the algebra A_2 is called the kernel of the homomorphism β. The kernel J_β of the homomorphism is a two-sided ideal. If the kernel is zero, the mapping is said to be a faithful homomorphism.

A faithful homomorphism of A_1 <u>onto</u> A_2 is an <u>isomorphism</u>.
A homomorphism of A_1 <u>into</u> A_1 is an <u>endomorphism</u>.
A homomorphism of A_1 <u>onto</u> A_1 is an <u>automorphism</u>.

A <u>representation</u> of an algebra is a homomorphism of the algebra A into the algebra of linear operators on a certain vector space. Further on we consider only $*$-homomorphisms and $*$-representations. Every algebra may be considered as a representation of itself. Take as the vector space on which the representation is constructed, the algebra A itself. With each element $a \in A$ corresponds the application $\bar{a}\, x$ of the vector space by defining

$$\bar{a}\, x = a\, x .$$

$a \to \bar{a}$ is a faithful representation if A is an algebra with unit. This representation is called the regular representation.

A function applying the algebra A into the real numbers $\mathbb{R}$, $a \rightsquigarrow \|a\|$ is called a norm on A if

(1) $\|a\| \geqslant 0$ and $\|a\| = 0$ if $a = 0$

(2) $\|a + b\| \leqslant \|a\| + \|b\|$

(3) $\|\lambda a\| = |\lambda|\ \|a\|$, $\lambda \in \mathbb{C}$.

If furthermore

(4) $\|ab\| \leqslant \|a\|\ \|b\|$

(5) If A contains a unit e, then $\|e\| = 1$

then A is called a normed algebra.

Using the norm $\|.\|$, we can introduce a metric in A, defining the distance by

$$d(a,b) = \|a - b\|,$$

and we can introduce the notion of convergence in this metric. A normed algebra, which is closed with respect to its norm is called a complete normed algebra. A norm *-algebra is a *-algebra with norm such that $\|a^*\| = \|a$ for all $a \in A$. If furthermore A is complete in this norm, A is called an involutive Banach algebra. A C^*-algebra is a Banach algebra such that

$$\|a^* a\| = \|a\|^2$$

for all $a \in A$.

We shall see later that every C^*-algebra may be represented (faithfully) as a subalgebra of L(H), that is the algebra of bounded operators on a certain Hilbert space.

2.2 States and Representations

A representation of the algebra A is any homomorphism of A into the algebra of linear operators on some vector space; this vector space is called the representation space. We restrict ourselves to $*$-representations of C^*-algebras. One example of a C^*-algebra is the set L(H) of all bounded linear operators on a given Hilbert space H. The algebra L(H) will play an important role in the sequel. By a $*$-representation of a C^*-algebra A we always mean a $*$-homomorphism of $a \to \pi(a)$ of A into the algebra L(H) for some H. Such a representation is always continuous. A representation is said to be cyclic if there exists a vector e in H such that the set of all vectors $\pi(a)e$, $a \in A$, is dense in H. The vector e itself is then said to be cyclic for the representation $a \to \pi(a)$. Two representations $a \to \pi(a)$, $a \to \pi'(a)$ in the spaces H and H′ respectively, are said to be equivalent if there exists an isometric mapping of H onto H′ under which the operator $\pi(a)$ is mapped into $\pi'(a)$ for all $a \in A$. In other words, if U is this isometric mapping and if $x' = Ux$, then $\pi'(a)x' = U\pi(a)x$. Consequently $\pi'(a)Ux = U\pi(a)x$ for all vectors x in H, i.e.

$$\pi'(a)U = U\pi(a).$$

A subspace $H_1 \subset H$ is said to be invariant if every vector in H_1 is mapped again into a vector in H_1 by all operators $\pi(a)$. If H_1 is a closed invariant subspace, then all the operators $\pi(a)$ can be considered as operators in H_1. We then obtain a representation in the space H_1. This representation is called a subrepresentation of the initial representation.

A linear functional on an algebra A is a mapping f of the algebra A into the complex numbers $\mathbb{C}$ such that

$$f(\alpha a + \beta B) = \alpha f(a) + \beta f(b) .$$

In this definition one only uses the vector space structure of the algebra A.

A continuous linear functional f on a normed algebra A is such that

$$|f(a)| \leqslant c \, \|a\|$$

for all $a \in A$ and some positive constant c. The smallest constant is denoted by $\|f\|$ and is called the norm of f. Functionals may be added in the usual way. One sees easily that

$$\|f\| + \|g\| \geqslant \|f + g\|$$

$$\|\alpha f\| = |\alpha| \, \|f\|,$$

and $\|f\| = 0$ implies $f = 0$ and vice versa.

With this norm the set of all continuous functionals on A becomes a Banach space; it is the dual space of A and is indicated by A^*. The norm induces a topology on A^*. Later on we construct another topology on A.

A linear functional on a C^*-algebra A is called positive if $f(a^*a) \geqslant 0$ for all $a \in A$.

We indicate some useful properties for positive linear functionals. Positive functionals satisfy the Schwartz inequality

$$|f(ab)|^2 \leqslant f(a^*a) \, f(b^*b) \quad .$$

Let A be a C^*-algebra with identity. Every positive linear functional on A is bounded and continuous, i.e.

$$|f(a)| \leqslant f(e) \, . \, \|a\| \qquad \text{and} \qquad \|f\| = f(e) \, .$$

A state on a C^*-algebra A is a positive linear functional such that $f(e) = 1$.

A state is pure if it cannot be written as a convex sum of other states. Thus a pure state is extremal in the convex set S of all states on A.

We now introduce another topology on A^*, the dual space of A. Every fixed $a \in A$ defines a functional on A^* by $h_a(f) = f(a)$. They are part of A^{**}, the space of continuous functionals on A^*. The weak *-topology on A^* is defined as the weakest topology, i.e. with the smallest possible assembly of open sets, in which every element of A is continuous, considered as a functional on A^*. This topology is Hausdorff (i.e. separating) and is characterized by the sets of neighbourhoods

$$V_\varepsilon(f_o, M) = \{f \mid f \in A^*, \ |f(a) - f_o(a)| < \varepsilon, \text{ for all } a \in M\}$$

where M is a subset of A, containing a finite number of elements. The mathematical importance of this topology lies in the fact that A_1^* is compact in this topology, where A_1^* is the unit sphere of A^*. Because S is a closed subset of A^*, it is also compact.

2.3 The Gelfand, Naimark, Segal Construction

The Gelfand, Naimark, Segal construction associates to every positive, linear functional on a C^*-algebra A a representation of A in a certain $L(H)$. The idea is to consider A as a vector space and to use the left regular representation of A on this space. In order to define a scalar product, one uses the positive linear functional. The elements with norm zero form a subspace, by which one takes the quotient. Finally the pre-Hilbert space obtained is completed.

Let A be a C^*-algebra with identity e and f a positive linear functional on A. We define a hermitian form on A (considered as a vector space) by

$$(y, x) = f(y^*x) \qquad \text{for } x, y \in A.$$

The set J_f of all elements $x \in A$ such that $(x, x) = 0$ form a left ideal of A. Indeed, suppose $y \in A$, then

$$|(yx, yx)|^2 = |f((yx)^* yx)|^2 = |f((x^*y^*y)\, x)|^2 \leqslant$$
$$\leqslant f((x^*y^*y)^* \, (x^*y^*y))\, f(x^*x) = 0$$

using the Schwartz inequality. Thus $yx \in J_f$. One verifies easily that J_f is a vector space.

Let us denote by H' the quotient space A/J_f, i.e. the set of all classes of elements of A, equivalent modulo the elements of J_f. In every class $\rho_f(x)$, $\rho_f(y)$, ... we choose an element x, y as a representative of its class. We define in H' a scalar product

$$(\rho_f(y), \rho_f(x)) = (y, x) = f(y^* x).$$

This scalar product does not depend upon the particular representatives we have chosen. Indeed, take x, x' in the class $\rho_f(x)$, then

$$|(y, x-x')|^2 = |f(y^*(x-x'))|^2 \leqslant$$
$$\leqslant f(yy^*) \, f((x-x')^* (x-x')) = 0$$

because $x-x' \in J_f$. It is easy to see that $(\rho_f(y), \rho_f(x))$ is a scalar product. The completion of H' with respect to this inner product will be denoted by H; H is a hilbert space. We now construct a representation of A in the space H. Let $a \in A$. We define a representation π_f of A by

$$\pi_f(a) \, \rho_f(x) = \rho_f(ax) \quad .$$

The operator $\pi_f(a)$ is bounded. Indeed

$$\| \pi_f(a) \, \rho_f(x) \|^2 = (\pi_f(a) \, \rho_f(x), \pi_f(a) \, \rho_f(x)) =$$
$$= (a\, x, a\, x) = f(x^* a^* a\, x).$$

We want to use known inequalities. We put therefore

$$f_1(a) = f(x^* a\, x).$$

Clearly f_1 is a positive linear functional, so that

$$f_1(a^* a) = f(x^* a^* a\, x) = f((a\, x)^* a\, x) \geqslant 0 \; .$$

Consequently, A being a C^*-algebra with identity

$$\|f_1(b)\| \leqslant f_1(e) \; \|b\|,$$

i.e. for $b = a^* a$

$$f_1(a^* a) \leqslant f_1(e) \; \|a\|^2$$

thus

$$f(x^* a^* a\, x) \leqslant f(x^* x)\ \|a\|^2$$
$$= (\rho_f(x), \rho_f(x))\ \|a\|^2$$

and consequently

$$\|\pi_f(a)\ \rho_f(x)\|^2 \leqslant \|a\|^2\ \|\rho_f(x)\|^2 .$$

Thus $\pi_f(a)$ is bounded on H' and we have

$$\|\pi_f(a)\| \leqslant \|a\| .$$

But H' is dense in H by construction, and therefore this operator can be uniquely extended to a bounded operator in H. In this connection, the norm $\|\pi_f(a)\|$ does not change and the equality written above remains valid for the norm of the operator $\pi_f(a)$ in the space H. The mapping $a \to \pi_f(a)$ is a representation of the algebra A. Indeed

$$\pi_f(\alpha a + \beta b)\rho_f(x) = \rho_f((\alpha a + \beta b)x)$$
$$= \rho_f(\alpha a\ x) + \rho_f(\beta b\ x)$$
$$= \alpha\,\pi_f(a)\ \rho_f(x) + \beta\,\pi_f(b)\ \rho_f(x).$$

Also $\pi_f(ab)\ \rho_f(x) = \rho_f(ab\ x) = \pi_f(a)\ \rho_f(bx)$

$$= \pi_f(a)\ \pi_f(b)\ \rho_f(x) .$$

π_f is a *-homomorphism: $\pi_f(a^*) = \pi_f(a)^*$. Indeed

$$(\rho_f(y), \pi_f(a)\ \rho_f(x)) = f(y^* a\ x) = f((a^* y)^*\ x)$$
$$= (\rho_f(a^* y), \rho_f(x)) = (\pi_f(a^*)\ \rho_f(y), \rho_f(x)).$$

The vector $\rho_f(e)$ is a cyclic vector, because

$$\pi_f(A)\,\rho_f(e) = H'$$

and H' is dense in H.

A representation $\pi(A)$ of an algebra A in L(H) is said to be irreducible if $\pi(A') = \{\lambda I\}$, where $\pi(A)'$ is the set of operators in L(H), commuting with $\pi(A)$. If π is induced by a positive, linear functional, then $\pi(A)$ is irreducible if and only if f is a pure state on A.

2.4 Automorphisms of a C^*-algebra.

Let us be given a representation π of a C^*-algebra A.

An automorphism τ of A is called internal in the representation π if it is implemented by a unitary operator; if this is not possible it is called external.

A state f on a C^*-algebra is called invariant for an automorphism τ if

$$f(\tau[a]) = f(a) \quad \text{for all} \quad a \in A.$$

Theorem. If a state f is invariant for an automorphism τ, then the automorphism is internal and

$$\pi_f(\tau[a]) = U_\tau \, \pi_f(a) \, U_\tau^{-1}$$

and

$$U_\tau \rho_f(e) = \rho_f(e) .$$

i.e. the cyclic vector is an eigenvector of $U(\tau)$, the unitary operator implementing the automorphism

Proof. Let τ be the automorphism of A such that $\tau[A] = A$. Let f be a state on A, invariant for the automorphism τ, i.e.

$$f(\tau[a]) = f(a) \quad \text{for all} \quad a \in A.$$

We prove that τ is unitarily implementable. Using the G.N.S. construction for the state f, we construct the Hilbert space H such that

$$f(a) = (\rho_f(e), \pi_f(a) \rho_f(e)) .$$

The automorphism introduces a unitary operator on H.

Define

$$U_\tau \rho_f(x) = \rho_f(\tau[x]) .$$

Because f is invariant for τ, this relation defines a linear operator on H in a consistent manner; the subdivision into equivalence classes of A is not destroyed. It is a unitary operator because

$$(U_\tau \rho_f(y), U_\tau \rho_f(x)) = (\rho_f(\tau[y]), \rho_f(\tau[x]))$$
$$= f(\tau[y]^*, \tau[x]) = f(\tau[y^*x]) = f(y^*x)$$
$$= (\rho_f(y), \rho_f(x)) .$$

This unitary operator implements τ. Indeed

$$(\rho_f(y), \pi_f(\tau[a]) \rho_f(x)) = f(y^*\tau[a] x)$$
$$= f(\tau^{-1}[y^*\tau[a] x]) = f(\tau^{-1}[y]^* a \tau^{-1}[x])$$
$$= (\rho_f(\tau^{-1}[y]), \pi_f(a) \rho_f(\tau^{-1}[x]))$$
$$= (U_\tau^{-1} \rho_f(y), \pi_f(a) U_\tau^{-1} \rho_f(x))$$
$$= (\rho_f(y), U_\tau \pi_f(a) U_\tau^{-1} \rho_f(x)).$$

Remark: The C^*-algebra A will later on play the role of the algebra of quasi-local observables of a physical system and the states f will be considered as quantum-mechanical states. We shall make the fundamental supposition that the dynamics produces an automorphism of the algebra A. If there exists a state f, which is invariant for the dynamical automorphisms τ_t, then the dynamics is unitarily implementable in the corresponding representation. If further on the situation is such that for this particular state f the expression $f(a\,\tau_t[b])$ is continuous in t for all $a, b \in A$, then the Stone theorem tells us that this one-parameter group is continuous and that there exists an infinitesimal generator, which may be identified with the Hamilton operator. Here the special choice of the representation plays a fundamental role.

2.5 Von Neumann Algebras

Let H be a separable Hilbert space, and let $L(H)$ be the set of all bounded linear operators on H. We have seen before that $L(H)$ is a C^*-algebra.

Let M be some subset of $L(H)$. We denote by M' the set of all elements of $L(H)$ commuting with all elements of M. M' is called the commutant of M. It is easy to see that M' is an algebra, and that this algebra always contains the multiples of the identity. In the same way one defines

$$M'' = (M')',$$

the bicommutant, and it is clear, that $M'' \supset M$. Further, if $M_1 \subset M_2$, then $M_1' \supset M_2'$. From this property one deduces that

$$M' = M''' = M^{(5)} = \ldots$$

and

$$M'' = M^{(4)} = M^{(6)} = \ldots$$

If $M = M^*$, M' is a *-algebra.

A <u>von Neumann algebra</u> M on H is a *-subalgebra of $L(H)$ such that $M = M''$.

The centre Z of a von Neumann algebra is $M \cap M'$. It is also the centre of M'.

A von Neumann algebra, whose centre only contains the multiples of the identity, is called a <u>factor</u>.

Factors will turn out to be very important in the physical applications.

If M is abelian, then clearly $M \subset M'$. M is called <u>maximal abelian</u> if $M = M'$.

Let A be a *-subalgebra of $L(H)$. A vector $x \in H$

is called a <u>cyclic vector</u> for A if

$$\{A\,x \mid x \in A\}$$

is dense in H. The vector x is called a <u>separating vector</u> if the condition $a \in A$, $a\,x = 0$ implies $a = 0$.

<u>Proposition 1</u>. Let S be a subset of H, A a *-sub algebra of L(H). Let P be the projection on the closure of the set $AS = \{a\,x \mid a \in A,\ x \in S\}$. Then $P \in A'$ and is the smallest projection of A' containing the whole set S.

<u>Theorem 2</u>. Let M be a von Neumann algebra. Then the following conditions are equivalent

(i) x is a cyclic vector for M.

(ii) x is a separating vector for M'.

The <u>weak operator topology</u> is generated by the set of neighbourhoods of an operator t defined by

$$V_\varepsilon(t, S_1, S_2) = \{a \mid a \in L(H),\ |(x, (t-a)\,y)| < \varepsilon\}$$

for all $x \in S_1$ and all $y \in S_2$, where S_1 and S_2 are finite subsets of H.

<u>Theorem 3</u>. A von Neumann algebra M is closed in the weak operator topology.

<u>Proposition 4</u>. Let A be a *-subalgebra of L(H) f a linear positive functional majorized by ω_x, where $\omega_x(a) = (x,\ a\,x)$. Then there exists a $T \in A'$ such that $f = \omega_{T\,x}$.

3. KINEMATICAL PROPERTIES OF PHYSICAL SYSTEMS

3.1 Construction of the Algebra of Quasi-local Observables

The algebra for an infinite system is constructed from the local algebras, corresponding to bounded regions

V of the space R^ν, (or Z^ν for a lattice system), where ν stands for the dimension of the space.

The local algebra, denoted by A(V), describes operations on a system, which are performed in the region V. Since the system is then finite, ordinary quantum mechanics provides us with an algebra, suitable for the system at hand. We illustrate the construction of the algebra for the infinite system by an example. Therefore we take an infinite system of Fermi particles. A system of fermions is described by creation and annihilation operators $\psi^*(\vec{x})$ and $\psi(\vec{x})$ $(\vec{x} \in V)$, acting irreducibly on a standard Hilbert space $H_F(V)$, called Fock space. These operators satisfy the canonical anticommutation relations:

$$\{\psi(\vec{x}), \psi^*(\vec{x}')\} = \delta(\vec{x} - \vec{x}')$$
$$\{\psi(\vec{x}), \psi(\vec{x}')\} = \{\psi^*(\vec{x}), \psi^*(\vec{x}')\} = 0.$$

In order to avoid the use of distributions, one introduces the operators $\psi^*(f)$ and $\psi(f)$, formally defined for every $f \in L^2(V)$ by

$$\psi^*(f) = \int_V \psi^*(\vec{x})\, f(\vec{x})\, d\vec{x} ,$$

and

$$\psi(f) = \int_V \psi(\vec{x})\, \overline{f(\vec{x})}\, d\vec{x} .$$

The anticommutation relations then take the following form

$$\{\psi(f), \psi^*(g)\} = (f, g)$$

and

$$\{\psi(f), \psi(g)\} = 0 .$$

These are bounded operators.

A(V) is the von Neumann algebra generated by all bounded operators on $H_F(V)$ of the form $\psi^*(f)\ \psi(g)$

In Fock space $H_F(V)$ there is one vector Ω, called the vacuum, with the property

$$\psi(f)\,\Omega = 0$$

for all $f \in L^2(V)$. Fock space is generated from this vacuum by repeated application of the operators $\psi^*(f)$.

The algebra $A(V)$ has the following properties, which we shall impose for the systems we consider.

1°. $A(V)$ is a C^*-algebra. This is clear since $A(V)$ is a von Neumann algebra.

2°. Isotony. $A(V_1) \subset A(V_2)$ if $V_1 \subset V_2$ and $A(V_1 \cup V_2) = \{A(V_1), A(V_2)\}$ (i.e. the C^*-algebra generated by $A(V_1)$ and $A(V_2)$.)

3°. Locality. $[A(V_1), A(V_2)] = 0$ if $V_1 \cap V_2 = \emptyset$.

We now consider representations of the local algebra $A(V)$. We only consider systems with a finite density, i.e. with a finite number of particles in V.

We are therefore only interested in representations for which the number operator

$$N_V = \int_V \psi^*(\vec{x})\,\psi(\vec{x})\,d\vec{x}$$

exists. We call such representations physical representations. For such representations we have the following property.

4°. All physical representations are locally normal, i.e. the representation is a direct sum of copies of the original irreducible representation and is therefore primary (factor) of type I.

We now proceed with the construction of the algebra for the infinite system.

Due to the isotony property the union of all $A(V)$ is again a $*$-algebra. When we complete this algebra in the norm, induced from the local algebras, we arrive at the C^*-algebra A of all quasi-local observables:

$$A = \overline{\bigcup_{V \in R^\nu} A(V)}^{\,norm}$$

The elements of this algebra are called quasi-local since they are norm limits of local observables. This has the consequence that truly global quantities like the total number operator are not contained in A. We reject them because their outcomes for a proper infinite system can only be infinite. This is the reason why representations of A are no longer copies of the Fock representation.

It can be shown that under rather general conditions the physical relevant representations are of type III.

In closing this section we prove the following lemma.

Lemma 1. For all physical representations π one has

$$\pi(A(V_0))' \cap \pi(A)'' = (A(\neg V_0))''$$

in which $A(\neg V_0)$ is defined as

$$A(\neg V_0) = \overline{\bigcup_{V} A(V)}^{\,norm} \quad \text{with} \quad V \cap V_0 = \emptyset$$

i.e. $\neg V_0$ is the complementary volume of V_0 in R^ν.

Proof. Due to the definition of $A(\neg V_0)$ and A we have

$$A = \{ A(\neg V_0), A(V_0) \} \quad .$$

In the representation space this implies

$$\pi(A)'' = (\pi(A(\neg V_0))'' \cup \pi(A(V_0))'')''.$$

Since $\pi(A(V_0))$ is a factor of type I (property 4^0), we can factorize the representation and the representatio space as

$$H = H_1 \otimes H_2$$

and

$$\pi(A(V_0))'' = L(H_1) \otimes 1.$$

Using the locality we see that $\pi(A(\neg V_0))'' \subset \pi(A(V_0))'$; in the decomposition this takes the form:

$$\pi(A(\neg V_0))'' = 1 \otimes L$$

where L is a weakly closed algebra on H_2. Substituting this in the formula for $\pi(A)''$ one finds:

$$\pi(A)'' = L(H_1) \otimes L.$$

For the commutant we get

$$\pi(A)' = (L(H_1 \otimes L)' = 1 \otimes L'.$$

We can write now

$$\begin{aligned} \pi(A(V_0))' \cap \pi(A)'' &= (\pi(A(V_0))'' \cup \pi(A)')' \\ &= (L(H_1) \otimes 1 \cup 1 \otimes L')' \\ &= (L(H_1) \otimes L')' \\ &= 1 \otimes L'' = \pi(A(\neg V_o))'' \end{aligned}$$

since L is weakly closed. This proves the lemma.

3.2 Automorphism Groups of the Algebra

In this section we introduce the concept of symmetry group acting on the system.

We start again by giving concrete examples. We are mainly interested in the three dimensional Euclidean group E^3 ($\nu = 3$) and the group of translations in time T. We consider first the Euclidean group.

The action of an element $g \in E^3$ on the algebra of the Fermi-system is defined on its generators by

$$\alpha_g[\psi(f)] = \psi(f_g); \quad f_g(\vec{x}) = f(g.\vec{x})$$
$$\alpha_g[\psi^+(f)] = \psi^+(f_g).$$

It is easily checked that this definition can be extended by continuity to a $*$-automorphism of the whole algebra. A $*$-automorphism of a C^*-algebra is automatically norm preserving, so we have:

$$\|\alpha_g[a]\| = \|a\|, \qquad a \in A.$$

This definition furnishes us moreover with the following structure:

$$\alpha_{g_1} \cdot \alpha_{g_2} = \alpha_{g_1 \cdot g_2}$$

and

$$\lim_{g \to g_0} \|\alpha_g[a] - \alpha_{g_0}[a]\| = 0 .$$

i.e. that α is a norm continuous representation of E^3 as a $*$-automorphism of A.

In the case of time evolution one would like to arrive at a similar situation by considering the formal definition:

$$\alpha_t[a] = e^{itH} a e^{-itH}$$

in which H stands for the total Hamiltonian of the system. Since this operator does not belong to the algebra of quasi-local observables it is not clear whether $\alpha_t[a]$ belongs again to A. We assume this to be true for the system we consider.

In order to keep the discussion as general as possible we consider in the sequel continuous $*$-representations of an arbitrary amenable group G.

Let us therefore discuss the concept of amenable group. Let G be a locally compact group and let C(G) denote the C^*-algebra of continuous, complex valued and bounded functions on G with sup-norm.

In particular $C(G)$ possesses an identity δ, namely the function which takes the value 1 for all $g \in G$. The group structure in G defines for every $g_0 \in G$ in a natural way two automorphisms of $C(G)$, denoted by $\alpha^L_{g_0}$ and $\alpha^R_{g_0}$ respectively and defined as

$$(\alpha^L_{g_0}\phi)(g) = \phi(g_0 g), \quad g_0 \in G, \ \phi \in C(G)$$

$$(\alpha^R_{g_0}\phi)(g) = \phi(g g_0), \quad g_0 \in G, \ \phi \in C(G).$$

A normalized linear functional on $C(G)$ is called an invariant mean if it is positive and invariant for $\alpha^L_{g_0}$ and $\alpha^R_{g_0}$ for all $g_0 \in G$. We denote the mean of $\phi \in C(G)$ by $\eta\phi$ or by $\eta\phi(\hat{g})$, in which $\hat{g}$ stands for a dummy variable.

A locally compact group G is called <u>amenable</u> when there exists an invariant mean on $C(G)$.

Every compact group is amenable since one can take the Haar measure as a mean.

Every abelian group is amenable. The Euclidean group is amenable since it is an extension of an amenable (abelian) group with an amenable (compact) group. The invariant mean is in general not unique on $C(G)$; one has however an unique value of η on a smaller class of functions, namely on the functions of positive type (i.e. the functions for which the matrices with elements $\phi(g_i^{-1} g_j)$ $i, j = 1, \ldots, N$ are positive). These functions and complex linear combinations of them correspond with the functions ϕ of the form

$$\phi(g) = (\phi, U(g)\psi)$$

for some unitary representation of G. We denote this class of functions by $N(G)$.

In the case of translations (in space or time) the unique value of a mean η on $N(G)$ takes the form

$$\eta(\phi) = \lim_{L\to\infty} \frac{1}{2L}\int_{-L}^{L} \phi(x)\, dx, \quad \phi \in N(R) .$$

In general we have for an amenable and separable group G that

$$\eta(\phi) = \lim_{n\to\infty} I_n(\phi); \quad \phi \in N(G)$$

in which I_n denotes an integral over a compact subset C_n of G. We arrive now at the notion of asymptotic abelianess. The importance of this notion resides in the fact that it enables us to extend the results of classical (abelian) ergodic theory to the quantum case.

In its strongest form the condition reads:

$$\lim_{g\to\infty} \left\| \left[\alpha_g [a] , b \right] \right\| = 0$$

for all $a, b \in A$ and g an element of a non compact group G. We refer to this property as norm-asymptotic abelianess.

This property follows from the construction of A if G is the space translation group or the Euclidean group.

The group representation is called weakly asymptotic abelian if

$$\lim_{g\to\infty} f(\left[\alpha_g [a] , b\right]) = 0$$

for all $f \in A^*$ and all $a, b \in A$.

The group representation is called η-asymptotic abelian if

$$\eta\, f\, (\left[\alpha_g [a], b\right]) = 0$$

for all $f \in A^*$ and all a, $b \in A$.

The different properties listed here are related in the following way:

norm-asymptotic ab. $\rightarrow$ weak-asymptotic ab. $\Rightarrow$ η-asymptotic ab.

The first implication follows from the fact that f is norm continuous, the second implication is a consequence of the fact that invariant means of functions, vanishing at infinity, are zero.

The asymptotic abelianess of translations in time however is in general an open problem to be decided from case to case. We explicitly mention when we make use of this condition in the sequel.

3.3 States on the Algebra of Quasi-local Observables

The expectation values of the measurements corresponding to a certain element of the algebra form a positive, linear and normalized functional on the algebra. These functionals will therefore be called states; the set of all states is denoted by S.

This structure is well known from ordinary quantum mechanics where the relevant states are of the form:

$$f(a) = (\psi, a\psi); \quad \psi \in H .$$

These states are called vector states

Statistical mechanics provides us with a more general example of state, namely

$$f_V(a) = \mathrm{Tr}\,(\rho_V\, a) .$$

Here ρ_V is a density matrix, which is a positive, trace-class operator with unit trace.

States that can be written in this form on Fock space are called here <u>normal states</u>.

Among the equilibrium states these examples are exhaustive for the finite system. This follows from the fact that one assumes existence of a number operator for the finite system.

Specializing to grand canonical equilibrium states, the density matrix has the form

$$\rho_V = \exp\left[-\beta(H_V - \mu N_V)\right] / \mathrm{Tr}\exp\left[-\beta(H_V - \mu N_V)\right]$$

In this expression H_V is the Hamilton operator and N_V the number operator for a system enclosed in a box with volume V; β stands for $(kT)^{-1}$ and μ is the chemical potential. Under suitable conditions the limit

$$f(a) = \lim_{V\to\infty} \mathrm{Tr}\,(\rho_V\, a), \quad a \in \bigcup_V A(V)$$

will exist. For infinite systems we always consider states of this form. The state is defined on A by continuous extension. This state, describing the infinite system, is no longer a normal state. When we consider the restriction of this state to a local algebra A(V), one has of course a normal state on A(V). This normal state is described by a density matrix $\hat{\rho}_V$, which is in general not exactly the same as the original density matrix ρ_V, due to the interaction with the surroundings. For T = 0 and given Hamiltonian (in fact we take $H_V - \mu N_V$), the ground state is supposed to exist as a vector in $H_F^V : \phi_o^V$.

Then we can form

$$g_V(a) = (\phi_o^V, a\,\phi_o^V), \quad a \in A(V)$$

and take the limit $V \to \infty$ of $g_V(a)$. By so doing

we obtain the ground state functional for the infinite system

$$g(a) = \lim g_V(a) \quad .$$

States which are normal when restricted to a local algebra are called <u>locally normal states.</u> We shall consider only such states.

It can easily be deduced from their definition that the states f and g are invariant in time. The equilibrium state is furthermore invariant for the Euclidean group when the Hamiltonian is Euclidean invariant; this we always assume. We have then

$$f(a) = f(\alpha_g[a]) = f(\alpha_t[a]) \text{ for } a \in A; \quad g \in E^3; \; t \in T.$$

A <u>physical system</u> is defined as an algebra of observables, symmetry groups acting on it and the convex space of states on this algebra.

A special role will be played in the sequel by the ergodic invariant states. Let us first recall some notions of classical ergodic theory. In this case observables are functions on a space Γ. The evolution in time maps the space Γ into itself and invariant states correspond to measures on Γ, which are invariant under this map. A measure μ is called ergodic if Γ cannot be decomposed into invariant subsets, i.e. all subsets of Γ invariant for T have measure zero or have a complement with measure zero. The importance of this definition resides in the fact that for ergodic measures the mean of a certain observable in time equals the mean with respect to the measure μ (mean over T-space). One expects that ergodic measures correspond to pure phases.

In the non-commutative theory, that we are discussing now, ergodic states correspond with the extremal points of the convex, compact set S_0 of all time invariant states. These states have the property that they can be decomposed only trivially into invariant states; we have therefore a natural extension of the notion of ergodic measure. The equilibrium states for the infinite system as defined above need not be ergodic in general, due to the occurrence of phase transitions in nature. We shall see however that for asymptotic abelian systems, every invariant state can uniquely be decomposed into ergodic states. Physically this decomposition amounts to a reconstruction of the pure phases occurring in a mixture.

This decomposition can also break the Euclidean symmetry that the equilibrium state originally possessed. One has then the possibility to describe crystals. One of the main objectives of these notes will be the classification of the possible symmetries that the pure phases inherit from the equilibrium state.

Among the states that are invariant for a certain symmetry a special place is again taken by the extremal points of this set, which will be called extremal invariant states; we reserve the word ergodic for states extremal invariant in time.

3.4 Cluster Properties

Just as in the classical ergodic theory there is an intimate connection between ergodicity (extremal invariance) and cluster properties.

A state f is called weakly clustering for the amenable group G acting on A if

$$\eta f(\alpha_{\hat{g}}[a]\, b) = \eta\, f(\alpha_{\hat{g}}[a])\, f(b)$$

for all a, b $\in$ A. For states invariant for G this reduces to

$$\eta f(\alpha_{\hat{g}}[a]\, b) = f(a)\, f(b)\ .$$

We shall prove that for G-asymptotic abelian systems, an invariant state is extremal invariant if and only if it is weakly clustering.

Also stronger forms of the cluster property will be used in the sequel. We mention the following ones: An invariant state is called weakly mixing if

$$\eta \left| f(\alpha_{\hat{g}}[a]\, b) - f(a)\, f(b) \right| = 0 \quad \text{for all } a, b \in A.$$

For the translation group in three-dimensional space it is possible to introduce a cluster property lying in between weak clustering and weak mixing. We call this partial weak mixing

With the obvious notation a possible formulation of partial weak mixing looks like

$$\eta^{X} \left| \eta^{Y,Z} f(\alpha_{\hat{a}}[a]\, b) - f(a)\, f(b) \right| = 0\ .$$

An invariant state is called strongly clustering if

$$\lim_{g\to\infty} f(\alpha_{g}[a]\, b) = f(a)\, f(b) \text{ for all } a, b \in A \text{ and } G \text{ non-compact.}$$

This property is well known to hold for correlation functions for gases and liquids when the arguments are moved apart.

Finally we introduce the notion of uniform

<u>clustering</u> which only applies when G stands for translations in space. A state is called uniformly clustering if for fixed $a \in A$ and all $b \in A$ $(\neg S(R))$ one has: ($S(R)$ is a sphere with radius R)

$$|f(b-a) - f(b)\,f(a)| \leqslant \varepsilon(R)\ \|b\|$$

with $\lim_{R \to \infty} \varepsilon(R) = 0$

We have the following relations between these cluster properties: uniform clustering $\Rightarrow$ strong clustering $\Rightarrow$ weak mixing $\Rightarrow$ partial weak mixing $\Rightarrow$ weak clustering.

3.5 Properties of the Mean of an Operator

Let us consider a state f on A. The G.N.S. representation furnishes us with a cyclic representation π_f of A in $L(H)$ with cyclic vector $\rho_f(e)$ such that

$$f(a) = (\rho_f(e), \pi_f(a)\,\rho_f(e)) \quad \text{for all } a \in A.$$

When the state f is invariant for the group G acting as an automorphism group on A we have moreover a strongly continuous unitary representation U_f of G in $L(H)$ such that

$$U_f(g)\,\pi_f(a)\,U_f(g)^{-1} = \pi_f(\alpha_g[a]) \text{ for all } a \in A \text{ and } g \in G$$

and

$$U_f(g)\,\rho_f(e) = \rho_f(e) \quad \text{for all } g \in G.$$

When it is clear which state we have in mind we omit the suffix f.

In the representation space we define the notion of <u>concrete mean of an operator</u> $\pi(a)$ over an amenable group. For every $a \in A$ we define $\eta_f(a) \in L(H)$ as

$$(x, \eta_f(a)\, y) = \eta(x, \pi(\alpha_{\hat{g}}[a])\, y) \quad \text{for all } x, y \in H.$$

This definition makes sense since the application

$$x, y \to \eta(x, \pi(\alpha_{\hat{g}}[a])\, y)$$

is a bounded sesquilinear form; by the theorem of Rietz there exists a bounded operator corresponding to the definition above.

The fact that the mean η is linear and positive implies directly that the map: $a \to \eta_f(a)$ is linear, preserves conjugation and that it maps positive elements into positive operators.

We now prove the following lemma.

<u>Lemma 1</u>. Let f be a G-invariant state. The covariance algebra R_f is defined as the von Neumann algebra generated by $\pi_f(a)$ and $U_f(G)$, i.e. $R_f = \{\pi_f(A) \cup U_f(G)\}''$.
Further we denote by E_0 the projection on the closed subspace of the vectors invariant for $U(G)$. Then

1^0 $\eta_f(a) \in \pi_f(A)'' \cap U_f(G)'$

2^0 $\eta_f(a)\, \rho_f(e) = E_0\, \pi_f(a)\, \rho_f(e)$

When G acts moreover η-asymptotically abelian, we have:

3^0 R_f' is abelian and is contained in $\pi_f(A)''$

4^0 The centre of R_f is contained in the center of $\pi_f(A)''$

5^0 $\eta_f(a)$ is an element of the centre of R_f.

<u>Proof</u>. The definition of η_f implies that $\eta_f(a)$ commutes with every element of $\pi_f(A)'$; hence $\eta_f(A) \in \pi_f(A)''$.

The fact that η is an invariant mean implies that $\eta_f(A)$ is invariant for $U_f(g)$; thus $\eta_f(A) \in \pi_f(A)'' \cap U_f(G)'$. We now prove the second statement. We have, for $y \in H$

$$(y, \eta_f(a)\,\rho_f(e)) = \eta\,(y, U_f(\hat{g})\,\pi_f(a)\,\rho_f(e))$$
$$= \eta\,(y, U_f(\hat{g})\,(E_o + (1 - E_o))\,\pi_f(a)\,\rho_f(e)).$$

Since E_o projects on the invariant vectors one has

$$U_f(g)\,E_o = E_o = E_o\,U_f(g)\,.$$

Consequently we have

$$(y, \eta_f(a)\,\rho_f(e)) = (y, E_o\,\pi_f(a)\,\rho_f(e))$$
$$+ \eta(y, (1 - E_o)\,U(\hat{g})\,\pi(a)\,\rho_f(e)) \quad .$$

By the definition of η_f we may write this as

$$(y, \eta_f(a)\,\rho_f(e)) = (y, E_o\,\pi_f(a)\,e_f)$$
$$+ (y, (1 - E_o)\,\eta_f(a)\,\rho_f(e)) \;.$$

But $\eta_f(a)$ is G-invariant; therefore the last term vanishes and we have

$$(y, \eta_f(a)\,\rho_f(e)) = (y, E_o\,\pi_f(a)\,\rho_f(e)).$$

Since this holds for all $y \in H$, the proof of 2° is complete. We now come to the case that G acts η-asymptotically abelian. Since the functional h defined by

$$h(a) = (x, \pi_f(a)\,y) \qquad x, y \in H$$

is an element of A^*, we have

$$\eta\,h\,([\alpha_{\hat{g}}\,[a]\;,\;b]) = 0 \qquad a, b \in A\;.$$

Using the definition of η_f this implies

$$(x, \eta_f(a)\,\pi_f(b)\,y) = (x, \pi_f(b)\,\eta_f(a)\,y).$$

Since this holds for all $x, y \in H$ we have: $\eta_f(a) \in \pi_f(A)'$. In combination with 1° this yields:

$$\eta_f(a) \in \pi_f(A)'' \cap \pi_f(A)' \cap U_f(G)$$

i.e. $\eta_f(a)$ is an invariant element of the centre of $\pi_f(A)''$.

Let us consider now the commutant of R_f; this can be written as

$$R_f' = \pi_f(A)' \cap U_f(G)' .$$

Let us take $b \in R_f'$; we want to show that $b \in \pi_f(A)''$. Since the representation is cyclic we can approximate the vector $b\,\rho_f(e)$ with a sequence $\pi_f(a_n)\,\rho_f(e)$, $a_n \in A$. It is then easy to show that $\eta_f(a_n)\,\rho_f(e)$ converges also to $b\,\rho_f(e)$.

Indeed $\|\pi_f(a_n)\rho_f(e) - b\,\rho_f(e)\| \leq \varepsilon$ implies

$$\| E_o\, \pi_f(a_n)\, \rho_f(e) - E_o b\, \rho_f(e)\| \leq \varepsilon .$$

By virtue of 2° we have $E_o \pi_f(a_n)\, \rho_f(e) = \eta_f(a_n)\, \rho_f(e)$ and since b is invariant for G we have $E_o b\, \rho_f(e) = b\, \rho_f(e)$. Hence

$$\|\eta_f(a_n)\, \rho_f(e) - b\, \rho_f(e)\| \leq \varepsilon .$$

When d is an arbitrary element of $\pi_f(A)$ we use the fact that both $\eta_f(a_n)$ and b are elements of $\pi_f(A)'$ to conclude

$$\|\eta_f(a_n)\, d\, \rho_f(e) - bd\, \rho_f(e)\| \leq \varepsilon\, \|d\|.$$

Since the vectors $d\, \rho_f(e)$ are dense in H, this implies that $\eta_f(a_n)$ converges strongly to b.

The operators $\eta_f(a_n)$ belong to $\pi_f(A)''$ which is weakly and hence strongly closed. Consequently b is contained in $\pi_f(A)''$ and we proved $R_f' \subset \pi_f(A)''$.

We have a fortiori $R_f' \subset (\pi_f(A) \cup U_f(G))'' = R_f$; hence R_f' is abelian. This completes the proof of 3°.

Next we consider the centre of R_f which is

$$R_f \cap R_f' = (\pi_f(A) \cup U_f(G))'' \cap R_f' .$$

Since $R_f' \subset \pi_f(A)''$ we have

$$R_f \cap R_f' = R_f' = R_f' \cap \pi(A)'' = \pi_f(A)' \cap \pi_f(A)' \cap U_f(G) .$$

This means that the elements of the centre of R_f are precisely the invariant elements of the centre of $\pi_f(A)''$. This proves 4°. Since we showed already that $\eta_f(a)$ is an invariant element of the centre of $\pi_f(A)''$, this proves also the last statement.

3.6 Extremal Invariant States

We now start the discussion of extremal invariant states. The following lemma characterizes extremal invariant states and η-weakly clustering states in terms of the corresponding representation.

Lemma 1. 1°) A G-invariant state f is extremal invariant if and only if R_f is irreducible.

2°) An arbitrary state f is weakly clustering with respect to G if and only if $\rho_f(e)$ is an eigenvector of $\eta_f(a)$ for all $a \in A$.

Proof. Suppose f is extremal invariant. Let P be a projection operator from R_f'; so P commutes with $\pi(A)''$ and is invariant for $U(g)$. The functional, defined by

$$g(a) = (P\rho(e), \pi(a)\rho(e))$$

is a positive functional majorized by f. (See also prop. 4, section 2.5). The functional g is invariant since P commutes with $U(g)$. Therefore we have the following decomposition of f into invariant and positive functionals.

$$f = g + (f - g).$$

Since f is extremal invariant this decomposition must be trivial, i.e.

$$g = \lambda f.$$

or

$$(P\rho(e), \pi(a)\rho(e)) = \lambda(\rho(e), \pi(a)\rho(e)) .$$

Since $\pi(a)\rho(e)$ is dense in H, this equation has as only solution $P = 0$ or $P = 1$. Hence the commutant of R_f consists of scalars and R_f acts irreducibly.

In order to prove the converse suppose now that R_f is irreducible, and let there be given a decomposition of f into invariant positive functionals: $f = g_1 + g_2$.

According to prop. 4, section 2.5, there exists an operator b in $\pi(A)'$ such that

$$g_1 = (b\rho(e), \pi(a)\rho(e)) .$$

The invariance of g_1 implies moreover that $b \in U(g)'$. Hence we have $b \in R_f'$. Since R_f is irreducible, b can only be a multiple of the identity; g_1 is therefore a multiple of f. This means that the decomposition is trivial and the proof of 1°) is complete.

We now prove the second part. Weak clustering implies that

$$\eta(\rho(e), \pi(\alpha_{\hat{g}}[a])\pi(b)\rho(e)) = \eta(\rho(e), \pi(\alpha_{\hat{g}}[a])\rho(e)) \cdot (\rho(e), \pi(a)\rho(e))$$

for all $a, b \in A$. Using the definition of η_f this is equivalent with

$$(\rho(e), \eta_f(a)\pi(b)\rho(e)) = (\rho(e), \eta_f(a)\rho(e)) \cdot (\rho(e), \pi(b)\rho(e))$$

or equivalently

$$((\eta_f(a) - (\rho(e), \eta_f(a)\rho(e))^* \rho(e), \pi(b)\rho(e)) = 0.$$

Since $\pi(b)\rho(e)$ is dense, this is true if and only if

$$\eta_f(a)^* \rho(e) = \overline{(\rho(e), \eta_f(a)\rho(e))}\ \rho(e) .$$

Since A is a *-algebra and since η_f preserves conjugation, this means that $\rho(e)$ is an eigenvector of all $\eta_f(a)$.

This completes the proof of the lemma.

We now come to the relation between extremal invariance and weak clustering.

Theorem 2. Let G be an amenable group, f a G-invariant state and $U(g)$ the unitary representation of G in H, then the following two conditions are equivalent

1° f is weakly clustering with respect to G.

2° E_0 is one-dimensional.

These conditions imply (and are equivalent with those for as. abelian G).

3° f is extremal G-invariant.

Proof. Suppose f is weakly clustering. According to lemma 1 this is equivalent with

$$\eta_f(a)\rho(e) = \lambda_a \rho(e) .$$

Following lemma 1 in the previous section we have

$$\eta_f(a)\rho(e) = E_0 \pi(a)\rho(e)$$

Hence $E_0 \pi(a)\rho(e) = \lambda_a \rho(e)$.

Since $\rho(e)$ is cyclic, this is equivalent with the condition that E_0 projects on the one-dimensional subspace generated by $\rho(e)$.

The extremal invariance of f can, due to lemma 1, be deduced from the irreducibility of R_f. Suppose b is an element of R_f'. The vector $b\rho(e)$ is then invariant since b commutes with $U(g)$. The fact that the subspace of invariant vectors is one-dimensional implies

$$b\,\rho(e) = \lambda_b\,\rho(e)\,.$$

But b is also an element of $\pi(A)'$, and since $\rho(e)$ is separating for this algebra (Theorem 2, section 2.5), we conclude that b is a scalar. Hence R_f acts irreducibly when E_o is one-dimensional. This concludes the proof of $1 \rightarrow 2 \Rightarrow 3$. The proof of $2 \Rightarrow 1$ is trivial.

We now prove $3 \Rightarrow 1$ for η-asymptotically abelian groups. By lemma 1 we have that extremal invariance implies that R_f is irreducible. The centre of R_f consists therefore of scalars. If we have an η-asymptotically abelian group, lemma 1 of the previous section assures that $\eta_f(a)$ is an element of the centre of R_f. Consequently $\eta_f(a)$ is a scalar and each vector, in particular $\rho(e)$, is an eigenvector of $\eta_f(a)$. By lemma 1 this implies that f is weakly clustering.

This closes the proof of the theorem.

The following theorem on the decomposition of a G-invariant state f into extremal G-invariant states can be proved.

<u>Theorem 3</u>. Let f be a G-invariant state on A, and let the amenable and separable group G act η-asymptotically abelian on A. Then there exists a Radon measure μ on the compact convex set S_o of all invariant states, representing f, i.e. such that

$$f(a) = \int_{S_o} g(a)\, d\mu(g)\,,$$

and with the property that every Baire set of S_o, that is disjoint from the extremal points of S_o, has measure zero. This measure μ is unique.

The proof is rather tedious and may be found elsewhere. (See f.e. 1) and 2)).

3.7 Cluster Properties and the Spectrum of U(g)

Up to now we only discussed the concept of weak clustering in the representation. The following two propositions give the relative strength of the remaining cluster properties in this framework.

We specialize now to abelian groups, isomorphic with R^ν, having in mind space or time translations.

By Stone's theorem there exists a projection valued measure E on R^ν such that a continuous and unitary representation U of G is decomposed as

$$U(\vec{a}) = \int_{R^\nu} \exp(i\vec{a}.\vec{p})\, dE(\vec{p}) \quad .$$

One of the results of theorem 1 of the previous section is that it provides a characterization of weak clustering in terms of the spectrum of U. It states that weak clustering is equivalent with the fact that $E(O) = E_0$ is one-dimensional. The following proposition gives analogous properties for weak mixing and strong clustering.

Proposition 1. One considers $G \simeq R^\nu$. Let f be a G-invariant state and let E_0 be one-dimensional (or equivalently let f be weakly clustering). Then

1^0 f is weakly mixing if and only if $\vec{p} = 0$ is the only discrete point of the spectrum of U_f.

2^0 f is strongly clustering when the spectral measure of U_f is, except for $\vec{p} = 0$, absolutely continuous with respect to the Lebesgue measure on R^ν.

We omit the rather technical proof of this proposition.

(See (1) or (2)).

As far as partial weak mixing is concerned, it may be proved that the discrete spectrum contains other points than $\vec{p} = 0$. They lie in these directions, in which the state is not weakly mixing but only weakly clustering.

The last cluster property, the uniform clustering, is of a somewhat different character. Its importance resides in the validity of the following proposition.

<u>Proposition 2</u>. Let f be an arbitrary locally normal state on A. The following two conditions are equivalent.

1° f is uniformly clustering with respect to space translations.

2° $\pi_f(A)''$ is a factor (or the state f is primary).

Here we also omit the proof. (See references 1 or 2). We notice that it is not only of mathematical interest to know when a state is primary since the primarity of a state implies that whenever this state is invariant for some η-asymptotic abelian group, it is also extremal invariant for that group. This is a simple consequence of lemma1, section 3.5, which implies that R_f' is trivial, whenever the center of $\pi_f(A)$ is trivial and of lemma 1, section 3.6, which states that the irreducibility of R_f implies that f is extremal invariant. The proposition above shows that one needs rather strong cluster properties in space in order to show that a state is primary, In the next chapter we show that for translations in time it suffices to assume only weak clustering in order to get a primary state.

4. DYNAMICAL PROPERTIES OF PHYSICAL SYSTEMS

4.1 The K.M.S. Condition

The way in which we arrived at the state for the infinite system in the previous chapter is as yet not quite satisfactory from a fundamental point of view. One has to rely still too heavily on the finite system in the limiting procedure.

There is however a possible alternative definition that does not have this drawback. This alternative makes use of the K.M.S. (Kubo - Martin - Schwinger) boundary condition. This property will play an important role in the sequel. For the finite system the K.M.S. condition is well known in Green function theory. For the formulation of this property in the grand canonical ensemble we make a small change in the definition of the time automorphism.

Let $H_V' = H_V - \mu N_V$, then we define

$$\alpha_t^V[a] = \exp(itH_V')\, a \exp(-itH_V') \quad \text{for} \quad a \in A(V).$$

This definition describes physically still the same time evolution since the physically relevant operators commute with N_V. Consider now the functions $F_{ab}(t)$ and $G_{ab}(t)$ defined by the following equations

$$F_{ab}(t) = f_V(b\, \alpha_t[a])$$

$$G_{ab}(t) = f_V(\alpha_t[a]\, b)$$

Take now $z = t + i\gamma$, then replacing t by z we obtain $F_{ab}(z)$ and $G_{ab}(z)$. It is easily seen that

$F_{ab}(z)$ is defined in the strip $0 < \gamma < \beta$ and $G_{ab}(z)$ is defined in the strip $-\beta < \gamma < 0$. These functions are analytic in the strip, and furthermore continuous at the boundaries, in such a way that $F_{ab}(t)$ and $G_{ab}(t)$ are the boundary values. From the invariance of the trace under cyclic permutation one gets the K.M.S. boundary condition

$$F_{ab}(t + i\beta) = G_{ab}(t)$$

or

$$\mathrm{Tr}(b\, \alpha^V_{t+i\beta}[a]\rho_V) = \mathrm{Tr}(\rho_V\, \alpha^V_t[a]\, b) .$$

For the passage to the infinite system it is convenient to use the functions F and G only for real times.

Let $g \in S$, where S is the set of C^∞-functions on the real line, rapidly decreasing at infinity.

For functions $g \in S$, the Fourier transform $\hat{g}$ is defined by

$$\hat{g}(\varepsilon) = \frac{1}{2\pi}\int \exp(i\varepsilon t)\, g(t)\, dt \quad .$$

Then $\hat{g}(\varepsilon)$ belongs to the class D of infinitely differentiable functions with compact support.

Define, with $z = t+i\gamma$

$$g(z) = \int \hat{g}(\varepsilon) \exp(-iz\varepsilon)\, d\varepsilon \quad .$$

Then $g(z)$ is analytic in the entire z-plane and $t^n\, g(t+i\gamma)$ is a bounded function of t for any fixed γ and positive n. Let us therefore multiply the equation

$$F_{ab}(t+i\beta) = G_{ab}(t)$$

with $g(t)$ and integrate. We can shift the integration in the left hand side within the analyticity domain of F_{ab} and obtain

$$\int_{-\infty}^{+\infty} g(t)\, F_{ab}(t+i\beta)\, dt = \int_{-\infty}^{+\infty} g(t-i\beta)\, f_V(b\, \alpha_t [a])\, dt$$

Now

$$\int_{-\infty}^{+\infty} g(t-i\beta)\, f_V(b\, \alpha_t [a])dt = \int_{-\infty}^{+\infty} g(t)\, f_V(\alpha_t [a]b)dt$$

for all $g \in S$ and $a, b \in A(v)$.

When the time isomorphism for the infinite system can be defined in a suitable manner as limit of the time isomorphism for the finite system, it can be proved that the K.M.S. condition extends to the state for the infinite system. In this situation we have also for the infinite system

$$\int_{-\infty}^{+\infty} g(t-i\beta)\, f(b\, \alpha_t [a])dt = \int_{-\infty}^{+\infty} g(t)\, f(\alpha_t [a]b)dt$$

for $g \in S$ and $a \in A$.

It is now an easy calculation to show that for the finite system, where one has a normal state, the density matrix is uniquely determined by the K,M.S. condition and the evolution in time. The conjecture is then that the same is true for the infinite system. One tentatively defines then, given the automorphism α_t and the inverse temperature β, the equilibrium state as the locally normal state on A satisfying the K.M.S. condition. It is anyhow clear that, irrespective of the exact definition, the states of interest for the description of equilibrium are invariant states satisfying the K.M.S. condition. More information may be found in references 13) and 14)

4.2 General Properties of K.M.S. States

Lemma 1. Let $A_1 \equiv \pi_f(A)''$, $f_1(a) = (\rho_f(e), a\,\rho_f(e))$ with $a \in A_1$ and $\alpha_t'[a] = U_f(t)\, a\, U_f(t)^{-1}$, $a \in A_1$. Then f_1 satisfies the K.M.S. boundary condition with respect to A_1 and α_t', if f satisfies the same with respect to A and α_t.

Proof. We have to prove that

$$\int_{-\infty}^{+\infty} g(t-i\beta)(\rho_f(e), b\, U_f(t)\, a\, \rho_f(e))\, dt$$
$$= \int_{-\infty}^{+\infty} g(t)\, (\rho_f(e), a\, U_f^*(t)\, b\, \rho_f(e))\, dt$$

for $a, b \in \pi_f(A)''$, $g \in S$.

The proof is based on the fact that the original K.M.S. condition ensures the validity of this formula for $a_0, b_0 \in \pi_f(A)$. One proves that the integrals above are continuous for fixed b (resp. a) with respect to the weak topology, on the unit ball of $\pi_f(A)''$. By the density theorem of Kaplanski one concludes then that the equality of the two integrals extends from $\pi_f(A)$ to $\pi_f(A)''$.

Theorem 2. When f is a time invariant state satisfying the K.M.S. condition then the center of $\pi_f(A)''$ is elementwise invariant for $U_f(T)$.

Proof. Since $U_f(t)\, R\, U_f(t)^{-1} = R$ holds for $R = \pi_f(A)$, it holds for $R = \pi_f(A)'$ and hence for $R = \pi_f(A)''$, and therefore $R = Z$, the centre of $\pi_f(A)''$.

Consequently we have

$$(\rho_f(e), b\, U_f(t)\, c\, \rho_f(e)) = (\rho_f(e), c\, U_f^*(t)\, b\, \rho_f(e))$$

for all $b \in \pi_f(A)''$ and $c \in Z$.

In terms of the spectral decomposition of $U_f(t)$

with spectral measure $E(\omega)$ we have

$$(\rho_f(e),\, b\, U_f(t)\, c\, \rho_f(e)) = \int e^{i\omega t}\, d\mu(\omega)$$

$$\mu(\omega) = (b^* \rho_f(e),\, E(\omega)\, c\, \rho_f(e)) .$$

The K.M.S. condition then takes the form

$$\int \hat{g}(\varepsilon) \exp(-i\varepsilon(t-i\beta)) \exp(i\omega t)\, d\mu(\omega)\, d\varepsilon\, dt$$

$$= \int \hat{g}(\varepsilon) \exp(-i\varepsilon t) \exp(i\omega t)\, d\mu(\omega)\, d\varepsilon\, dt \quad .$$

Since $\hat{g}$ is a function of the class D we can perform first the integration over ω and t; this yields

$$\int \hat{g}(\omega)(1 - \exp(-\omega\beta))\, d\mu(\omega) = 0 .$$

Since this must hold for all $\hat{g}$ in D, we conclude that the measure μ is concentrated in $\omega = 0$.

Consequently we have

$$(\rho_f(e),\, b\, U_f(t)\, c\, \rho_f(e)) = (\rho_f(e),\, b\, c\, \rho_f(e)) .$$

Since this is true for all $b \in \pi_f(A)''$ and since c is in the centre of $\pi_f(A)''$ this implies

$$U_f(t)\, c\, U_f(t)^{-1} = c, \qquad c \in Z.$$

This completes the proof of the theorem.

Corollary. Let f be a state invariant for time translations satisfying the K.M.S. condition.

1° When f is extremal invariant (or η-weakly clustering) then f is primary.

2° When time translations act moreover η-asymptotically abelian we have:

a) The centre of $\pi_f(A)''$ coincides with the centre of R_f.

b) The three conditions of 1° are equivalent.

Proof. The centre of $\pi_f(A)''$ is elementwise invariant for $U_f(T)$ i.e.

$$\pi_f(A)'' \cap \pi_f(A)' \cap U_f(T)' = \pi_f(A)'' \cap \pi_f(A)' .$$

Hence we conclude that the centre of $\pi_f(A)''$ is contained in

$$(\pi(A)'' \cup U(T))'' \cap \pi(A)' \cap U(T)'$$

i.e. contained in the centre of R_f. We know that extremal invariance implies that R_f' is trivial. In particular this implies that its centre is trivial, consequently the centre of $\pi_f(A)''$ is also trivial. Because weak clustering implies extremal invariance this ends the proof of 1°.

When time translation acts η-asymptotically abelian, we proved that the centre of R_f is contained in the centre of $\pi_f(A)''$. Together with the converse statement, which was a consequence of the K.M.S. condition, this proves 2°. 2^b is an immediate consequence of the general theorem on extremal invariant states.

We recall that, because f is time invariant, the G.N.S. construction yields an operator $U_f(t)$, implementing the time automorphism. It may be proved that $U_f(t)$ may be written in the form

$$U_f(t) = \exp(iH_f t)$$

where H_f is the Hamiltonian. This Hamiltonian has now a symmetric spectrum.

If $T = 0$, many of these properties remain true. The ground state is still stationary and there exists a unitary operator $U_f(t)$ such that $U_f(t)\, \rho_f(e) = \rho_f(e)$, and $U_f(t) = \exp(iH_f t)$, but now the lowest eigenvalue of H_f is zero. The function $f(\alpha_t[a]\, b)$ can now be continued in the whole lower half-plane and $f(b\, \alpha_t[a])$ in the whole upper half-plane.

The same conjecture may now be made as the one we

made for the finite temperature case: All states, which are stationary and have positive spectrum, are possible candidates for ground states of the system.

More information may be found in reference 15).

5. SYMMETRY BREAKING

5.1 Transitive States

We start this section by showing that states arising from the decomposition into extremal invariant states, are still physically acceptable equilibrium states, i.e. are invariant in time and satisfy the K.M.S. condition. It will be useful to introduce the symbols η^T, η^S and η^E for resp. the mean over translations in time, space translations and over the Euclidean group; we denote further by S_o^T, S_o^S, S_o^E the states invariant in time, space and for the Euclidean group.

Lemma 1. Suppose T, R^3 and E^3 act η-asymptotically abelian on A. Let f be a K.M.S. state, invariant in time and invariant for E^3. Let μ be any of the representing Radon measures concentrated on respectively the extremal points of S_o^T, S_o^S or S_o^E in the sense of theorem 2, section 3.6, then the support of μ consists of states invariant in time and satisfying the K.M.S. condition.

The proof of this lemma may be found elsewhere. (See references 1) and 2)).

We see from this lemma that the decomposition of the original equilibrium state into extremal Euclidean invariant states yields again equilibrium (K.M.S.) states with the full Euclidean symmetry.

Physically this decomposition corresponds with a decomposition into states which correspond with a pure phase or with a mixture of only one phase (e.g. crystals shifted in space, a superposition of magnetic systems with different orientations of the magnetisation, etc.), whereas the original state can of course be a mixture of different phases (gas-liquid). The symmetry breaking that we want to analyse occurs when one decomposes the so obtained extremal Euclidean invariant state further into ergodic states. We consider therefore in the sequel the class P of all states on A which are invariant in time, extremal Euclidean invariant and satisfying the K.M.S. condition.

It is of course possible that an extremal Euclidean invariant equilibrium state represents already a pure phase, i.e. is extremal invariant in time. We call this type of states SP_1. Their characterization is a direct consequence of the results of Chapters III and IV.

Theorem 1. Suppose E^3 and T act η-asymptotically abelian. A state $f \in P$ is of type SP_1 if and only if f is uniformly clustering in space.

Proof. f is uniformly clustering if and only if f is primary. Since f is a K.M.S. state the primarity of f is equivalent with ergodicity.

Following Kastler (16) we introduce the notions of transitivity and the related concepts of orbit and stabilizer.

The orbit (for the Euclidean group) of a state g is the set O_g^E of states defined by

$$O_g^E = \{ \alpha_h^t g ; \ h \in E^3 \}$$

in which α_h^t is the transposed of α_h to A^*.
The <u>stabilizer</u> is defined as the subset H_g^E of E^3 which leaves the state g invariant, i.e.

$$H_g^E = \{ h \in E^3 \mid \alpha_h^t g = g \} .$$

H_g^E is a closed subgroup of E^3.

A state $f \in P$ will be called <u>transitive</u> if the measure μ that decomposes f into ergodic states is concentrated on one orbit, i.e. there exists a state g such that $\mu(O_g^E) = 1$. The states occurring in the decomposition of a transitive state represent physically the same pure phase since they are obtained from one physically pure state by an Euclidean movement.

The stabilizer H_g^E is the symmetry group of g. It is clear that the stabilizer of the state $\alpha_h^t g$ is $h^{-1} H_g^E h$, which means that the states in one orbit have conjugate symmetries. We denote by H_g^S <u>the closed subgroup of translations in</u> H_g^E. In the sequel we are especially interested in those cases in which H_g^S contains at least three non coplanar translations or equivalently for which R^3/H_g^S is compact.

Since O^3 is compact it is easy to see that the compactness of R^3/H_g^S implies the compactness of E^3/H_g^E; the converse however is not always true.

The following theorem shows that transitivity is equivalent with the compactness of E^3/H_g^E.

<u>Theorem 2</u>. Suppose E^3 and T act η-asymptotically abelian. Let $f \in P$ and let μ denote the measure that decomposes f into ergodic states. Then the transitivity of f implies that E^3/H_g^E is compact for all $g \in \mathrm{Supp}(\mu)$. When A is separable the

compactness of E^3/H^E_g for all $g \in \mathrm{Supp}(\mu)$ implies that f is transitive.

The highly non trivial proof may be found in references 1) and 2).

We now come to a second important consequence of the transitivity of the measure μ. In general the support of a measure is a closed set which is not a Baire set. It can happen therefore that the support of a measure concentrated in Baire sense on the extremal points, does not contain extremal points. The following theorem shows that this does not happen when the measure is transitive; all states occurring in the decomposition are then ergodic states, which is very reassuring from the physical point of view.

Theorem 3. Suppose T acts η-asymptotically abelian. Let $f \in P$ be transitive and let μ denote the representing measure which is concentrated on the extremal points of S^T_o in Baire sense. Then the support of μ consists entirely of ergodic states.

The proof may be found in references 1) and 2).

We can summarize our results in the following way. A transitive state $f \in P$ yields in the decomposition states which are ergodic and which have symmetries that are all conjugate to one subgroup H^E_g of E^3; moreover $E^3/H^E_{g_o}$ is compact. Since this does not yet imply that $R^3/H^S_{g_o}$ is compact, the class of transitive states is still somewhat too large for our purposes. We introduce therefore the concept of strong transitivity. Let O^S_g denote the orbit of a state g under the action of the translations in space alone. We call a state f strongly transitive if f is transitive and if O_g^S is closed for some $g \in \mathrm{Supp}(\mu)$.

We have the following theorem.

Theorem 4. Let T act η-asymptotically abelian. A transitive state f is strongly transitive if and only if R^3/H_g^S is compact for all $g \in \mathrm{Supp}(\mu)$.

The proof may be found once again in references 1) and 2).

Since our main interest lies in ergodic states with at least three non coplanar translations in their symmetry group (i.e. R^3/H^S is compact) we can restrict by this theorem our investigation to strongly transitive states. We denote the subclass of P consisting of all strongly transitive states by SP.

Let us now indicate the different possibilities for the symmetries occurring in the decomposition of an SP state. The first possibility was discussed already at the beginning of this section, namely that an SP state is already ergodic such that its underlying symmetry is the full Euclidean symmetry. We have seen that states of this type, denoted by SP_1, are exactly the uniformly clustering states.

A second possibility is that the rotation symmetry is broken in the decomposition, but that $H^S = R^3$, i.e. the ergodic states possess still the full translation symmetry. We denote this type of states by SP_2.

In the third place it could happen that H^S is continuous in one or two directions and discrete in the remaining independent directions. Rotation symmetry is then certainly broken. This type of states is denoted by SP_3.

Finally there is the possibility that H^S is discrete in three independent directions. Since H^S is closed and since R^3/H^S is compact, the only possibility is then that H^S is generated by three non-coplanar translations, i.e. H^S is a lattice. This type of state is physically very important because the full symmetry of the ergodic states H^E is then a crystallographic group.

This last type of SP state is denoted by SP_4.

5.2 Strongly Transitive States

The four types of SP states may be characterized by different cluster properties.

Theorem 1. An SP state which is not an SP_1 state is not extremal invariant for translations.

Proof. See references 1) or 2).

We notice that all SP states are weakly clustering with respect to the Euclidean group, since they are extremal invariant for this group. When one restricts the mean over the correlations to space translations, only the states of type SP_1 keep their cluster properties by the theorem above. In fact one sees that an SP state is (with respect to translations) or clustering in the strongest sense or not clustering at all. In particular it follows that Euclidean invariant states describing crystals are not weakly clustering in space which means that correlations show long range order. This fact is important since it is one of the keystones of Landau's argument for the non existance of a critical point for the fluid-solid phase transition.

When an SP state is not an SP_1 state we can, by the theorem above, make a non trivial decomposition of this state into extremal invariant states. It turns out that the so obtained extremal translation invariant states obey cluster properties of different strength, which serve to characterize the different types of SP states by cluster properties in space. The motivation of doing this is that one gets in this way a rather constructive characterization of SP states in terms of the group R^3, whose action is much better known than that of T.

<u>Theorem 2</u>. Suppose T acts η-asymptotically abelian. R^3 acts weak asymptotically abelian. Let f be an SP state which is not extremal invariant with respect to translations and let ν denote the measure that decomposes f into extremal translation invariant states. Let g_0 be a state in the support of ν.

1° f is an SP_2 state if and only if g_0 is weak mixing.

2° f is an SP_3 state if and only if g_0 is partial weak mixing, but not weak mixing.

3° f is an SP_4 state if and only if g_0 is weakly clustering, but not partial weak mixing.

All cluster properties are with respect to space translations.

<u>Proof</u>. See references 1) or 2).

We remark that the more the symmetry is broken, the weaker the corresponding cluster property becomes. The classification of SP states, different from SP_1, is based on the spectral properties of the unitary operators representing translations in the representation corresponding with a state g_0.

REFERENCES

General

1) H.J.F. Knops, "Ergodic states and symmetry breaking in phase transitions. A C*-algebra approach." Thesis, Nijmegen 1969. Ed. H. Gianotten, Tilburg.

2) E.J. Verboven, "Introduction to the algebraic approach of quantum statistical mechanics," North-Holland Publ. Co. (To appear).

3) D. Ruelle, Cargese Lectures in Theoretical Physics, (1965) F. Lurcat.

4) N.M. Hugenholtz, "Fundamental Problems in Statistical Mechanics", II., North-Holland Publ. Co., Amsterdam 1968.

5) G. Emch, H. Knops, E. Verboven, Commun. Math. Phys. 8, 300 (1968).

6) D. Ruelle, "Statistical Mechanics", W.A. Benjamin, Inc., New York. To be published.

Section 2

2) Contains the proofs of the theorems stated.

7) J. Dixmier, "Les C*-algebras et leurs répresentations," Gauthier-Villars, Paris 1964.

8) J. Dixmier, "Les algebres d'opérateurs dans l'espace hilbertien," Gauthier-Villars, Paris 1957.

9) Naimark, "Normed rings", Noordhoff, Groningen 1959.

Section 3

10) S. Doplicher, R.V. Kadison, D. Kastler and D.W. Robinson, Comm. Math. Phys. 6, 101 (1967).

11) S. Doplicher, D. Kastler and E. Størmer, "Invariant states and asymptotic abelianess", preprint, Marseille 1968.

12) Kastler and Robinson, Commun. Math. Phys. 3, 151 (1966)

Section 4

13) R. Haag, N.M. Hugenholtz and M. Winnink, Commun. Math. Phys. 5, 215 (1967).

14) M. Winnink, "An application of C^*-algebras to quantum statistical mechanics of systems in equilibrium." Thesis, Groningen 1968.

15) H. Araki and H. Miyata, "On K.M.S. boundary condition," preprint, Kyoto University 1968.

Section 5

16) D. Kastler, R. Haag and L. Michel," Central decompositions of ergodic states," preprint, Marseille 1967.

17) H.J.F. Knops, Commun. Math. Phys. 12, 36 (1969).

THE MATHEMATICAL STRUCTURE OF THE BCS-MODEL AND RELATED MODELS

W. Thirring

CERN - GENEVA

1. INTRODUCTION

In systems with infinitely many degrees of freedom one encounters mathematical problems which one is not used to in elementary quantum mechanics. The BCS model is a non-trivial and not too unrealistic example exhibiting features which are typical for many field theoretic problems. In these lectures I shall comment on several of these mathematical questions. Since a good deal of the background material has been covered by previous lectures, I shall not talk about the physical motivations behind the BCS-model and will restrict the mathematical preparation to a discussion of infinite tensor products and traces. Also, I shall not derive this theory in the classical style of deducing lemmas and theorems but only try to convey the ideas behind it. There are many papers in which one can find the necessary theorems but few which explain the heuristics of it.

With these tools I will analyze later the BCS

model and related models and we will find that methods beyond classical mathematics are actually indispensable to understand what is going on.

2. INFINITE TENSOR PRODUCTS

2.1 Heuristics

If one tries to generalize the scalar product for a finite tensor product for a system of N degrees of freedom,

$$\langle x | y \rangle = \prod_{i=1}^{N} (x_i | y_i), \tag{2.1}$$

$$| x \rangle = | x_1 \rangle \otimes | x_2 \rangle \otimes \dots \otimes | x_N \rangle ,$$

to the case of $N = \infty$ one faces immediately the problem of convergence of the product $\prod_1^\infty$. Even if one is dealing with normalized $|x_i\rangle, |y_i\rangle$ such that $|(x_i | y_i)| \leq 1$ the product need not converge. This would be the case only if all factors were positive. Indeed, for the absolute values there are only the alternatives

$$\begin{aligned} \text{I}: \quad & \prod_{i=1}^{\infty} |(x_i | y_i)| \rightarrow c > 0 \\ \text{II}: \quad & \prod_{i=1}^{\infty} |(x_i | y_i)| \rightarrow 0 \end{aligned} \tag{2.2}$$

whereas for II we also have $\prod_{i=1}^{\infty} (x_i | y_i) \rightarrow 0$.

In the case of I there is the possibility that the phases of $(x_i | y_i)$ spoil the convergence. For instance, if $(x_j | y_j) = e^{i\varphi_j}$, ($\varphi_j$ real), we have $\prod_{i=1}^{\infty} (x_i|y_i) = e^{i\sum_{j=0}^{\infty} \phi_j}$ and this converges if $\sum \phi_i$ does. To cope with this one adopts the convention

$$\begin{aligned} I_a \quad &: \quad \prod_{i=1}^{\infty} (x_i|y_i) = c \text{ if } \prod_{i=1}^{\infty} (x_i|y_i) \rightarrow c \\ I_b, II \quad &: \quad \prod_{i=0}^{\infty} (x_i|y_i) = 0 \text{ otherwise} \end{aligned} \tag{2.3}$$

I_b means that if $\prod_i |(x_i|y_i)|$ converges to a value $\neq 0$ but $\prod_i (x_i|y_i)$ does not converge, the scalar product is put equal to zero nevertheless. It turns out that with this definition of the scalar product one can consistently construct a Hilbert space from the linear combinations of these vectors in the usual fashion. Furthermore it turns out that I_a and I give equivalence relations between vectors in the sense that the convergence of $\prod_i (x_i|y_i)$ and $\prod_i (y_i|z_i)$ (or $\prod_i |(x_i|y_i)|$ and $\prod_i |(y_i|z_i)|$ implies the convergence of $\prod_i (x_i|z_i)$ (or $\prod_i |(x_i|z_i)|$ respectively). In these the infinite tensor product space decays into classes of mutually orthogonal subspaces. The whole space is called C.T.P. (complete tensor product) and classes of the type I_a, strong equivalence classes or I.C.T.P. (incomplete tensor product). Classes of the type I are called weak equivalence classes. One may picture this as follows:

```
          ┌ weak equiv. class ┌ I.C.T.P.
          │                   │ I.C.T.P.
          │                   │ I.C.T.P.
          │                   └    ⋮
C.T.P.   ─┤
          │ weak equiv. class ┌
          │        ⋮          └
          └
```

Of course this is schematic since there are innumerably many weak equivalence classes and each contains innumerably many I.C.T.P. However, if the individual factors are separable Hilbert spaces then the I.C.T.P's are too.

The relevant theorem of this theory states the following. If one takes the algebra $\left\{\bigotimes_i B_i\right\}''$ of the bounded operators B_i of the individual Hilbert spaces $|x_i\rangle$ (defined according to common sense in the C.T.P.) and their weak limits then they do not lead out of the I.C.T.P. and are there represented irreducibly. In other words the C.T.P. gives a highly reducible representation of the B_i's, the I.C.T.P. being the irreducible sub-spaces. It turns out that within one weak equivalence class the representations are essentially the same — otherwise different. This means that the complication with the phase factors does not lead to anything usefully new and we shall therefore not distinguish between the various I.C.T.P. within one weak equivalence class. Let me illustrate these statements by a simple familiar example.

2.2 The Tensor Product Representations of the Algebra of a Spin Chain

Let us consider the commutation relations

$$\left[\sigma_j^{(\alpha)}, \sigma_{j'}^{(\alpha')}\right] = 2\,i\,\delta_{jj'}\,\varepsilon_{\alpha\alpha'\alpha''}\,\sigma_j^{(\alpha'')}$$

where $j = 1,2$; $\alpha = 1,2,3$.

If we construct the complete tensor product

$$H = \bigotimes_{j=1}^{\infty} H_j$$

(all H_j being two-dimensional), then we obtain a representation π of these commutation relations by

$$\sigma_j^{(\alpha)} \rightsquigarrow 1 \otimes 1 \otimes \cdots \otimes \underbrace{\sigma^{(\alpha)}}_{j\text{-th place}} \otimes 1 \otimes \quad \ldots \in B^{\pi}$$

where $\sigma^{(\alpha)}$ is a usual Pauli matrix. This representation decomposes into irreducible sub-representations π_ζ again defined by

$$\sigma_j^{(\alpha)} \rightsquigarrow 1 \otimes 1 \otimes \cdots \otimes \sigma^{(\alpha)} \otimes 1 \otimes \cdots$$

but now the operator on the r.h.s. acts in an incomplete tensor product, labelled by the index ζ

$$H = \bigotimes_{j=1}^{\infty}{}^{\zeta}\; H_j \quad .$$

That these representations are irreducible follows from the fact that

$$\pi_\zeta(\{\sigma_j^{(\alpha)}\}) = \bigotimes_{i=1}^{\infty}{}^{\zeta} B(H_i) ,$$

by definition of $\bigotimes_{j=1}^{\infty}{}^{\zeta} B(H_i)$, (note that every element $\in B(H_i)$ can be written as a linear combination of the unit matrix and the three Pauli matrices) and this is just

$$B(\bigotimes_{i=1}^{\infty}{}^{\zeta} H_i) .$$

Therefore π as well as all π_ζ are of type I ($\pi = \bigoplus_\zeta \pi_\zeta$).

If two equivalence classes ζ, ζ' belong to the same weak equivalence class ζ_w, then they are equivalent since there exists a unitary operator

$U\left[(z_i)_{i=1,2,..}\right] \in B(\bigotimes_{i=1}^{\infty} H_i)$, $(|z_i| = 1 \; \forall \; i)$, which has the property that

$$U \pi(\sigma_j^{(\alpha)}) U^* = \pi(\sigma_j^{(\alpha)}) \quad \forall j \; \forall \alpha$$

but transforms $\bigotimes_{i=1}^{\infty}{}^{\zeta} H_i$ into $\bigotimes_{i=1}^{\infty}{}^{\zeta'} H_i$.

In fact in the natural way of choosing the basis they are even identical.

If ζ, ζ' do not belong to the same weak equivalence class, then π_ζ and $\pi_{\zeta'}$ are inequivalent. This fact will be proved later. From now on, we shall identify all equivalence classes belonging to the same weak equivalence class, because they give equivalent representations.

There are many more representations than the above-mentioned ones. Later on, we shall, for example, discuss the thermodynamic representations of the BCS model. But it should be said that, up to now, one cannot construct explicitly all possible representations. However to facilitate later treatment of the BCS model we conclude this section with a discussion of certain properties of the representations in our example.

(i) Automorphisms

One kind of automorphism is the following:

$$\sigma_j^{(\alpha)} \rightsquigarrow M_{(j)}^{\alpha\beta} \sigma_j^{(\beta)} \quad ,$$

where $M_{(j)}^{\alpha\beta}$ is an orthogonal (real) (3,3) matrix. Other automorphisms are, for example:

$$\sigma_j^{(\alpha)} \rightsquigarrow \sigma_{P(j)}^{(\alpha)}$$

P being a permutation of the natural numbers.

For all families of matrices $M_{(j)}$, there is always a unitary operator in the complete tensor product such that

$$U \sigma_j^{(\alpha)} U^* = M_{(j)}^{\alpha\beta} \sigma_j^{(\beta)} \quad .$$

But only if U does not lead out of the equivalence classes, is $U \in \pi(\{\sigma_j^{(\alpha)}\})''$. However, this is not the case in general.

Let us analyse the situation in an irreducible representation π_ξ. Consider the automorphism

$$\sigma_j^{\pm} \rightsquigarrow e^{\pm i\phi} \sigma_j^{\pm} \quad .$$

If ξ is determined by $\binom{1}{0} \otimes \binom{1}{0} \otimes \ldots$, then there exists a $U(\phi) \in B(\bigotimes_{i=1}^{\infty}{}^{\xi} H_i)$. The generator is the (unbounded) self-adjoint operator

$$\sum_{j=1}^{\infty} (\sigma_j^{(3)} - 1)$$

and

$$U(\phi) = \exp(i \frac{\phi}{2} \sum_{j=1}^{\infty} (\sigma_j^{(3)} - 1))$$

If $\mathfrak{f}$ does not contain $\binom{1}{0} \otimes \binom{1}{0} \otimes \ldots$, but another vector, say

$$\begin{pmatrix} a \\ b \end{pmatrix} \otimes \begin{pmatrix} a \\ b \end{pmatrix} \otimes \ldots \quad (a, b \neq 0) ,$$

then one could also write formally

$$U_{(\phi)} = \exp(i \frac{\phi}{2} \sum_{j=1}^{\infty} (\sigma_j^{(3)} - 1))$$

but neither the l.h.s. nor the r.h.s. makes sense, since $U(\phi) \notin B(\otimes^{\mathfrak{f}} H_i)$ (i.e. such a U can only be defined in a larger Hilbert space, e.g. the C.T.P. and $\sum(\sigma_j^{(3)} - 1)$ makes a vector of infinite norm out of every non-zero vector.

(ii) Isomorphisms of the representations

Let $\pi_{\mathfrak{f}}$, $\pi_{\mathfrak{f}'}$ be two representations, such that there exists an isometric operator U fulfilling

$$U \, \pi_{\mathfrak{f}}(\sigma_j^{(\alpha)}) = M_{(j)}^{\alpha\beta} \, \pi_{\mathfrak{f}'}(\sigma_j^{(\beta)}) \, U .$$

The two algebras $\pi_{\mathfrak{f}}(\{\sigma_j^{(\alpha)}\})''$ and $\pi_{\mathfrak{f}'}(\{\sigma_j^{(\alpha)}\})''$ are then spatially isomorphic. They are equivalent if $M_j = 1 \, \forall j$.

The two C^*-algebras $\pi_{\mathfrak{f}}(\{\sigma_j^{(\alpha)}\})$ (the norm closure of all finite linear combinations of <u>finite products</u> of $\pi_\gamma(\sigma_j^{(\alpha)})$'s) and $\pi_{\gamma'}(\{\sigma_j^{(\alpha)}\})$ are isomorphic, but generally this isomorphism cannot be extended to the weak closures. Take, for example,

$$\vec{s} = \lim_{N \to \infty} N^{-1} \sum_{j=1}^{N} \pi_\gamma(\vec{\sigma}_j) .$$

There are a lot of représentations where this limit exists (in the strong operator topology). Define $|\vec{n}\rangle \in \mathbb{C}^2$ (up to a phase factor) by

$$\vec{n}\,\vec{\sigma}\,|\vec{n}\rangle = |\vec{n}\rangle \qquad (\langle\vec{n}|\vec{n}\rangle = 1).$$

Then in all equivalence classes containing

$$|\vec{n}\rangle \otimes |\vec{n}\rangle \otimes \quad \ldots\ldots \quad ,$$

$\vec{s}$ exists. (Even in many more equivalence classes, but not in all.) A simple calculation yields

$$N^{-1} \sum_{j=1}^{N} \pi_{\vec{n}}(\vec{\sigma}_j) \to \vec{n}$$

($\pi_{\vec{n}}$ being the representation in one of the above equivalence classes), and this limit depends on the representation! If there were an isometric operator U transforming, for example,

$$U\,\vec{\sigma}_j = M\,\vec{\sigma}_j\,U \quad ,$$

then consequently

$$U\,\vec{n} = M\,\vec{n}\,U \quad ,$$

but $\vec{n}$ is a c-number. Hence this automorphism is incompatible with the weak and strong operator topology; it is only compatible with the norm topology.

Elaborating these ideas, one can easily show that $\pi_\xi \neq \pi_{\xi'}$ unless $\xi = \xi'$ (moderately weak equivalence classes).

(iii) Time development

This is an automorphism depending on t. The simplest case one can consider are spins in a homogeneous external magnetic field $\vec{B}$, which points in the z-direction. Then all spins rotate around this direction. The unitary operator $U(t) \in B^*$ is in suitable units

$$x_1 \otimes x_2 \otimes \ldots \rightsquigarrow e^{i\frac{t}{2}\sigma_1^{(3)}} x_1 \otimes e^{i\frac{t}{2}\sigma_2^{(3)}} x_2 \otimes \ldots$$

It describes this rotation, but generally leads out of the I.C.T.P. Consequently, the formal generator

$$\sum_{j=1}^{\infty} \sigma_j^{(3)}$$

does not exist at all (except for $\binom{1}{0} \otimes \binom{1}{0} \otimes \ldots$) and if a suitable c-number is subtracted. The group $\{U(t)\}$ is weakly measurable - in fact, its matrix elements are for all C_0-vectors of the type

$$\langle a \mid U(t) \mid b \rangle = \begin{cases} 1 & \text{if } t = 0 \\ 0 & \text{otherwise} \end{cases}$$

for t sufficiently small - but since the C.T.P. is not separable, Stone's theorem does not apply. (Only in separable Hilbert spaces does weak measurability imply weak continuity of unitary groups.)

Therefore one might argue that it is better

not to consider the Schrödinger picture, but the Heisenberg picture. Even so the situation is not much improved: there exists an automorphism

$$\vec{\sigma}_j \rightsquigarrow M(t)\,\vec{\sigma}_j \;\doteq\; \vec{\sigma}_j(t) \quad ,$$

but this automorphism cannot be extended to the weak closure of the algebra but only to the smallest C^*- algebra containing all $\vec{\sigma}_j$'s.

3. TRACES

For the treatment of systems at finite temperatures one works with traces since the thermal expectation value is given by

$$\langle A \rangle \;=\; \mathrm{Tr}\, Ae^{-\beta H} \,/\, \mathrm{Tr}\, e^{-\beta H} \quad , \tag{3.1}$$

Whereas for finite dimensional spaces there is no difficulty in defining the trace by

$$\mathrm{Tr}\, A \;=\; \sum_{i=1}^{N} \left(e_i \mid Ae_i\right) \quad , \tag{3.2}$$

the e_i being an orthogonal basis; for $N = \infty$ we run into severe difficulties even for bounded A, not only that $\sum_1^\infty$ may not converge, or not absolutely converge, so that it may assume any value in a different basis. Even absolute convergence in one basis does not guarantee the convergence in another basis.

Take

$$\begin{pmatrix} \begin{matrix} 0 & 1 \\ 1 & 0 \end{matrix} & & & \\ & \begin{matrix} 0 & \frac{1}{2} \\ \frac{1}{2} & 0 \end{matrix} & & \\ & & \begin{matrix} 0 & \frac{1}{3} \\ \frac{1}{3} & 0 \end{matrix} & \\ & & & \ddots \end{pmatrix} \tag{3.3}$$

for which the trace is absolutely convergent and zero. In another basis it looks like

$$\begin{pmatrix} \begin{matrix} 1 & \\ & -1 \end{matrix} & & & \\ & \begin{matrix} \frac{1}{2} & \\ & -\frac{1}{2} \end{matrix} & & \\ & & \begin{matrix} \frac{1}{3} & \\ & -\frac{1}{3} \end{matrix} & \\ & & & \ddots \end{pmatrix}$$

for which Tr is only conditionally convergent and may assume any value by a further change of basis.

It turns out that if $(x \mid Ax) \geq 0 \quad \forall x$ the result of (3.2) for $N = \infty$ is independent of the choice of the basis. Thus for positive operators the trace is well defined but may assume the value ∞ (as for $A = 1$, $N = \infty$). We get therefore a finite well defined trace for linear combinations of positive operators with finite trace (trace-class). It turns out that multiplication by a bounded operator does not spoil this property so that the trace-class is a two-sided ideal.

One may argue that the requirements are too

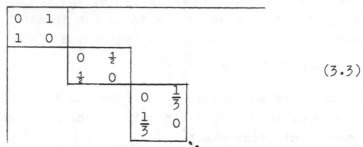

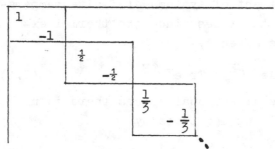

restrictive because the following may happen. Consider $H = C^{(2)} \otimes H_o$ where H_o is ∞-dimensional and consider the operators $\vec{\sigma} \otimes 1$. Since they are of the form

$$\begin{pmatrix} 1 & & & & & & \\ & -1 & & & & & \\ & & 1 & & & & \\ & & & -1 & & & \\ & & & & 1 & & \\ & & & & & -1 & \\ & & & & & & \ddots \end{pmatrix}$$

they do not satisfy the requirements and therefore one cannot define a trace for them. On the other hand, it is clear that this is a trivial situation and it would be silly not to define the trace just as for the σ's in $C^{(2)}$. To cope with this situation one needs to look more generally for a "relative trace" or "super trace" ϕ which is a positive linear functional over the positive operators of an algebra $\mathfrak{A}$ (it may also be infinite) satisfying

$$\text{(i)} \qquad \phi(\lambda A) = \lambda \phi(A) \qquad A \geqslant 0$$

$$\text{(ii)} \qquad \phi(A+B) = \phi(A) + \phi(B) \qquad (3.4)$$

$$\text{(iii)} \qquad \phi(UAU^{-1}) = \phi(A), \quad \forall\, U \text{ unitary}, \in \mathfrak{A}$$

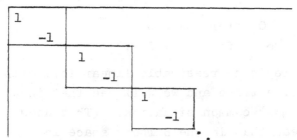

It turns out that these positive operators for which $\phi(A) < \infty$ are again the positive part of a two-sided ideal. What is more important, for a factor (which is an operator algebra $\mathfrak{A}$ such that the only operators $\in \mathfrak{A}$ commuting with all other operators of $\mathfrak{A}$, are a multiple of unity) ϕ is unique up to a positive number. Therefore one can use ϕ to classify factors. There are the following categories denoted by III, I, II_1 and II_∞.

III. The only possibility (except $\phi \equiv 0$) is

$$\phi(A) = \begin{cases} 0 & \text{for} \quad A = 0 \\ \infty & \text{for} \quad A \neq 0 \end{cases} .$$

In this case there is no reasonable linear form with the properties of a trace and we will see that in a way this is the most common situation. (This does of course not mean that in the Hilbert space in which the A's are acting one cannot define a trace but the trace-class has an empty intersection with our algebra.)

I. All bounded operators B in a Hilbert space are a factor (they are even irreducible) and because of the uniqueness of the trace ϕ must be of the usual form. These factors are denoted by I_N where N is the dimensionality of the Hilbert space. (Thus for I_∞ $\phi(1) = \infty$).

II. The remaining cases are denoted by II_1 or II_∞ depending on whether $\phi(1) < \infty$ or $\phi(1) = \infty$. We shall later encounter factors of type III and II_1, (II_∞ is always of the form $II_1 \otimes I_\infty$).

4. THE BCS-MODEL: GROUND STATE

The mathematical structure of the BCS model is rather simple, and one can treat it exactly in the case of a finite volume of a superconductor as well as in the infinite case.

For a finite volume, the hamiltonian is

$$H_{\Omega} = - \varepsilon \sum_{p=1}^{\Omega} (\sigma_p^{(z)} - 1) - \frac{2T_c}{\Omega} \sum_{p=1}^{\Omega} \sigma_p^- \sum_{p=1}^{\Omega} \sigma_p^+ \tag{4.1}$$

The $\sigma_p^{(\alpha)}$'s are Pauli matrices acting on the pth mode of a tensor product of two-dimensional Hilbert spaces

$$\mathcal{H} = \mathcal{H}_1 \otimes \mathcal{H}_2 \otimes \cdots \otimes \mathcal{H}_{\Omega}$$

with the dimension 2^{Ω}, where Ω is a number proportional to the volume. The $\vec{\sigma}_p$'s have nothing to do with the physical spin of the electrons, but σ_p^+ (σ_p^-) describes the creation (annihilation) of a Cooper pair, which has the kinetic energy 2ε.

We have made the assumption that the kinetic energy of all pairs is the same ("strong coupling limit"). Since there is only an interaction between pairs near the Fermi momentum, this is not completely unreasonable although not very realistic.

The model can be solved exactly if one introduces the operation

$$\vec{S} = \frac{1}{2} \sum_{p=1}^{\Omega} \vec{\sigma}_p \tag{4.2}$$

Then

$$H_{\Omega} = \varepsilon (\Omega - 2S_z) - \frac{2T_c}{\Omega} (\vec{S}^2 - S_z (S_z + 1)) \tag{4.3}$$

Defining r, N by

$$S = \frac{\Omega}{2} - r, \quad S_z = \frac{\Omega}{2} - N; \quad 0 \leq N \leq \Omega, \quad 0 \leq r \leq \text{Min}(N, \Omega - N) \tag{4.4}$$

(N is just the number of pairs), the eigenvalues of H_Ω are

$$E(N,r) = 2\varepsilon N + 2\,T_c(r-N)\left(1 + \frac{1}{\Omega}\right) + \frac{2T_c}{\Omega}\left(N^2 - r^2\right) \tag{4.5}$$

The structure of the lower lying levels is shown in fig. 1, where the energy of the ground state has been normalized to zero.

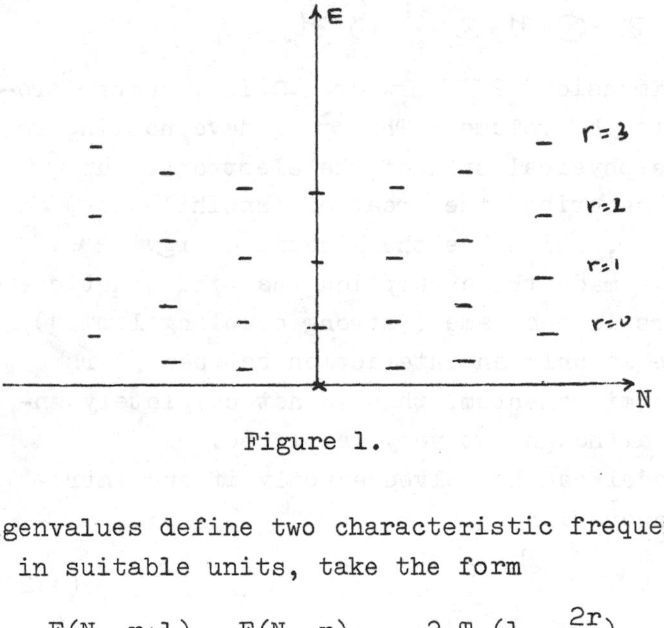

Figure 1.

The eigenvalues define two characteristic frequencies which, in suitable units, take the form

$$\Delta = E(N,\, r+1) - E(N,\, r) = 2\,T_c\left(1 - \frac{2r}{\Omega}\right)$$

$$-2\mu = E(N+1,\, r) - E(N,\, r) = 2\varepsilon - 2T_c\left(1 - \frac{2N}{\Omega}\right) \tag{4.6}$$

where μ is the chemical potential.

4.1 Time Development

$$\vec{\sigma}_p \rightsquigarrow \vec{\sigma}_p(t) = e^{iHt}\, \vec{\sigma}_p\, e^{-iHt} \tag{4.7}$$

If one defines

$$\vec{s} = \Omega^{-1} \sum_{p=1}^{\Omega} \vec{\sigma}_p , \tag{4.8}$$

then

$$s^{\pm}(t) = s^{\pm}(0)\, e^{\mp it(2\varepsilon - 2T_c s_z)} ,$$

$$s_z = \text{const.}, \quad \frac{d}{dt}(\vec{s}.\vec{\sigma}_p) = 0 . \tag{4.9}$$

Equation (4.9) shows that $\vec{s}$ rotates with a frequency 2μ around the z-axis, and one can deduce that the $\vec{\sigma}_p$'s rotate around $\vec{s}$ with a frequency Δ.

4.2 The Case $\Omega \to \infty$, Treatment "a la Haag"

We now consider what happens, if $\Omega \to \infty$, in the I.C.T.P. representations.

There are some representations, where

$$\vec{s}_\Omega = \Omega^{-1} \sum_{p=1}^{\Omega} \vec{\sigma}_p$$

converges towards a certain c-number $\vec{s}$. For example the representations characterized by

$$|\{\vec{n}\}\rangle \doteq |\vec{n}\rangle \otimes |\vec{n}\rangle \otimes \ldots \tag{4.10}$$

where $\vec{s}$ converges towards $\vec{n}$ and one might expect that $\vec{s} - \vec{n}$ can be neglected. (Note, however, that in the I.C.T.P. containing

$$\binom{1}{0} \otimes \binom{0}{1} \otimes \binom{0}{1} \otimes \binom{1}{0} \otimes \quad \cdots \quad \otimes \binom{0}{1} \otimes \cdots$$

4 times 8 times ...

this limit does not exist.)

In these representations, (4.10), Haag's method assumes that

$$H_{\Omega} = -\varepsilon \sum_{p=1}^{\Omega} \sigma_p^{(z)} - \frac{2T_c}{\Omega} \sum_{p,p'} (\sigma_p^{+} - n^{+})(\sigma_p^{-} - n^{-})$$

$$- 2T_c \sum_{p} (\sigma_p^{+} n^{-} + \sigma_p^{-} n^{+}) + \text{const.} \quad (4.11)$$

converges towards H_B (B stands for Bogoliubov) which is the linear part of H_{Ω} and has the form

$$H_B = \frac{\Delta}{2} \sum_{p=1}^{\infty} (\vec{n}\,\vec{\sigma}_p - 1). \quad (4.12)$$

One notices that $H_{\Omega} - H_B$ is quadratic in $\vec{s} - \vec{n}$ and that H_B and H_{Ω} have (if $\Omega \to \infty$) the same commutators with all σ_p's. Explicitly,

$$H_B = -\sum_{p=1}^{\infty} \left\{ \varepsilon\, \sigma_p^{(z)} + T_c \left(n^{(x)} \sigma_p^{(x)} + n^{(y)} \sigma_p^{(y)} \right) \right\} \quad (4.13)$$

in the I.C.T.P's containing $|\{\vec{n}\}\rangle$, and, comparing (4.12) with (4.13), one concludes that the "gap equation"

$$T_c = \Delta/2, \qquad n_z = \cos\Theta = \varepsilon/T_c \quad (4.14)$$

must hold, i.e. $\mu = 0$, since $n_z = \lim S_z/S$. (The point is that

$$\sum_{p=1}^{\infty} (\vec{n}\,'\vec{\sigma}_p - \text{const.})$$

can only be defined in the I.C.T.P.'s characterized by $|\{\vec{n}\}\rangle$ if $\vec{n}\,' = \vec{n}$, otherwise this operator makes a "vector" of infinite norm out of each non-zero vector.)

4.3 Explicit Treatment

One can check all assumptions explicitly and obtain the following results

(i) H_Ω (or a suitable c-number) converges <u>weakly</u> in the I.C.T.P.'s where

$$\lim \Omega^{-1} \sum_{p=1}^{\Omega} \vec{\sigma}_p = \vec{s} \qquad (4.15)$$

exists and the consistency relation (4.14) ("gap equation") is fulfilled which says physically that $\vec{s}$ does not change in time (i.e. there is no rotation around the z-axis). A ground state exists in the I.C.T.P. of $|\{\vec{n}\}\rangle$.

(ii) $e^{iH_\Omega t}\,\vec{\sigma}_p\,e^{-iH_\Omega t}$ converges strongly if $\vec{s}$ exists, and its limit is

$$e^{iH_B t}\,\vec{\sigma}_p\,e^{-iH_B t}$$

if $\vec{s}$ is constant in time. Otherwise the time development of $\vec{\sigma}_p$ consists of a rotation around $\vec{s}$ which itself rotates around the z-axis. In this case there is no Hamiltonian (obtainable either by limiting processes from the H_Ω's, or otherwise) governing the time evolution. For, although a $U(t)$

exists for this motion in the C.T.P ., it leaves no vector $\neq 0$ and therefore no I.C.T.P. invariant. Thus $U(t)$ is in none of these representations weakly continuous and can nowhere be written as e^{-iUt}. Furthermore H_B gives the wrong time dependence since it gives the rotation around $\vec{s}$ which is kept constant.

(iii) $e^{iH_\Omega t} \not\rightarrow e^{-iH_B t}$

since

$$\lim \langle\{\vec{n}\}| e^{iH_\Omega t} |\{\vec{n}\}\rangle = (1 + it\, T_c)^{-\frac{1}{2}} \tag{4.16}$$

but

$$\langle\{\vec{n}\}| e^{iH_B t} |\{\vec{n}\}\rangle = 1 .$$

(iv) One can formulate the problem in the language of spin waves, defined by

$$\vec{\sigma}_\lambda = \Omega^{-\frac{1}{2}} \sum_{p=1}^{\Omega} e^{i\lambda p} \vec{\sigma}_p , \tag{4.17}$$

and consider the limit $\Omega \rightarrow \infty$. One obtains a different result: in the representation where the gap equation holds, H_Ω converges strongly towards a self-adjoint operator H (on a dense domain) which is not equal to H_B.

$$H = T_c \left[\sum_{\lambda \neq 0} \tau_{-\lambda}^{-} \tau_\lambda^{+} + \left(1 - \frac{\omega^2}{T_c^2}\right) p^2 \right] \tag{4.18}$$

$$\tau_0^{\pm} = (q \pm ip)/\sqrt{2} , \quad \tau_{-\lambda}^{-} = (\tau_\lambda^{+})^{*}$$

The $\vec{\tau}$'s are rotated $\vec{\sigma}$'s. If $\Omega = \infty$, they obey pure boson commutation relations

$$[\tau^+_\lambda , \tau^-_{\lambda'}] = \delta_{\lambda,-\lambda'} \tag{4.19}$$

The spectrum of H is the same as for H_B for almost all degrees of freedom, only one degree of freedom ($\lambda = 0$) has a continuum in its part of the hamiltonian. (For a free particle this continuum corresponds to states with the same r and different N which, for $\Omega \to \infty$, become degenerate.) It is reminiscent of the rotation of $\vec{s}$ in the sense that in these representations $\vec{s}$ does not rotate but there is also no force which keeps $\vec{s}$ fixed.

5. THE BCS-MODEL: THERMODYNAMIC LIMIT

We shall now calculate the thermodynamic expectation values

$$\langle A \rangle_\Omega = \mathrm{tr}\left[e^{-H_\Omega/T} A\right] / \mathrm{tr}\left[e^{-H_\Omega/T}\right] \tag{5.1}$$

of elements A belonging to the *-algebra Σ' generated by the $\vec{\sigma}_p$'s. However, we are not really interested in the value of $\langle A \rangle_\Omega$ itself, but rather in the limit $\Omega \to \infty$ which we hope will exist. (The *-algebra Σ' is of infinite dimensions; there is a little problem in defining the trace but one should notice that Σ' consists only of finite linear combinations of elements of the form $\sigma_1^{(\alpha_1)} \ldots \sigma_n^{(\alpha_n)}$, etc., and for these $\langle A \rangle_\Omega$ is well-defined if $\Omega \gg n$. Hence, the question of the existence of $\lim \langle A \rangle_\Omega$ makes sense for all $A \in \Sigma'$.) The reader should bear in mind that in the expression defining $\langle A \rangle_\Omega$, S_z, and hence the number of pairs, is not kept fixed so that $\langle A \rangle_\Omega$

corresponds to the grand-canonical expectation value.

Our investigations on the ground state suggest the following conjecture. Splitting H_Ω into two parts,

$$
\begin{aligned}
H_\Omega &= H_{B,\Omega} + H'_\Omega \\
H_{B,\Omega} &= -\varepsilon \sum_{p=1}^{\Omega} \sigma_p^{(z)} - 2\,T_c \sum_{p=1}^{\Omega} \left(\sigma_p^+ \langle \sigma^- \rangle_B + \sigma_p^- \langle \sigma^+ \rangle_B\right) \\
&= -T\,\omega \sum_{p=1}^{\Omega} (\vec{\sigma}_p \vec{n}) \\
H'_\Omega &= -\frac{2T_c}{\Omega} \sum_{p=1}^{\Omega} \left(\sigma_p^+ - \langle \sigma^+ \rangle_B\right) \sum_{p'=1}^{\Omega} \left(\sigma_{p'}^- - \langle \sigma^- \rangle_B\right) \\
&\quad - 2\,T_c\,\Omega \langle \sigma^+ \rangle_B \langle \sigma^- \rangle_B
\end{aligned}
\tag{5.2}
$$

we know that in the ground state, $H'_\Omega \to 0$ and also does not contribute to the time development of the $\vec{\sigma}_p$'s when $\Omega \to \infty$. Our conjecture is now that H'_Ω also does not contribute to the thermodynamic expectation values when $\Omega \to \infty$, i.e.

$$\lim_{\Omega \to \infty} \langle A \rangle_{B,\Omega} = \lim_{\Omega \to \infty} \langle A \rangle_\Omega \tag{5.3}$$

with

$$\langle A \rangle_{B,\Omega} = \mathrm{tr}\left[e^{-H_{B,\Omega}/T} A\right] / \mathrm{tr}\left[e^{-H_{B,\Omega}/T}\right] \tag{5.4}$$

if the self-consistency relations ($\langle \vec{\sigma} \rangle_B = \vec{n}\,\mathrm{th}\,\omega$ where th denotes hyperbolic tangent)

$$\omega = \frac{T_c}{T}\,\mathrm{th}\,\omega\,, \quad \cos\theta = \varepsilon / T\omega \tag{5.5}$$

are fulfilled. They are the generalization of (4.14) to finite temperatures. But, certainly, the above-stated form of our conjecture cannot be true, since for example,

$$\langle \sigma^{(\alpha)} \rangle_{\Omega} = 0$$

$$\langle \sigma^{(\alpha)} \rangle_{B,\Omega} = n^{(x)} \operatorname{th} \omega \neq 0 \qquad (5.6)$$

in general. However, we shall show that the correct thermodynamic limit is a gauge invariant mean over the thermodynamic limit, which is obtained by using $H_{B,\Omega}$ instead of H_{Ω}

$$\lim_{\Omega \to \infty} \langle A \rangle_{\Omega} = \lim_{\Omega \to \infty} (2\pi)^{-1} \int_0^{2\pi} d\phi \langle A \rangle_{B,\Omega} \qquad (5.7)$$

(Our self-consistency conditions fix only the angle Θ, whilst ϕ is not determined.)

Before we go over to the proof, two remarks should be added:

i) the energy gap $T\omega$ depends on the temperature and becomes smaller with increasing temperature corresponding to the decrease of Δ with (r);

ii) if $T = T_c$ then $\omega = 0$.

Both facts can easily be seen in fig. 2.

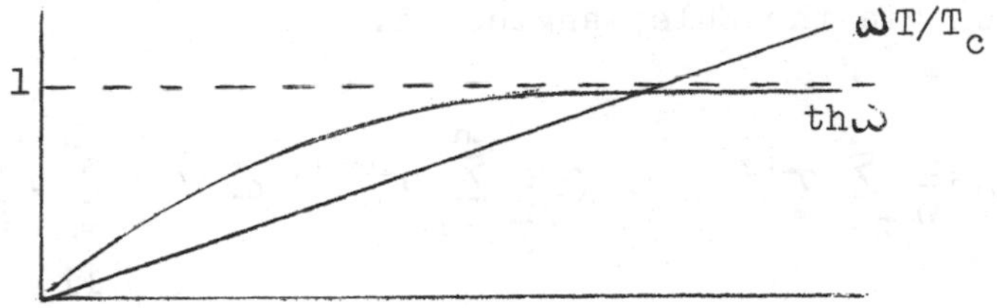

Figure 2.

Now let us indicate how one calculates the thermodynamic limit.

(a) The r.h.s. of (5.7).

We shall use a generating functional

$$\hat{A}_\Omega = \exp(i \sum_{p=1}^{\Omega} a_p \sigma_p^{(x)}) \exp(i \sum_{p=1}^{\Omega} b_p \sigma_p^{(y)}) \exp(i \sum_{p=1}^{\Omega} c_p \sigma_p^{(z)}) \quad . \tag{5.8}$$

By differentiating $\langle \hat{A}_\Omega \rangle_\Omega$, one obtains all expectation values. We can even use a simpler functional

$$A_\Omega = \exp(i \frac{a}{\Omega} \sum_{p=1}^{\Omega} \sigma_p^{(x)}) \exp(i \frac{b}{\Omega} \sum_{p=1}^{\Omega} \sigma_p^{(y)}) \exp(i \frac{c}{\Omega} \sum_{p=1}^{\Omega} \sigma_p^{(z)}) \tag{5.9}$$

because H_Ω as well as $H_{B,\Omega}$ are invariant under all permutations of the indices. Then, for example,

$$\langle \sigma_p^{(x)} \rangle = \frac{\partial}{\partial(a)} \langle A_\Omega \rangle_\Omega \Big|_{a = b = c = 0} \tag{5.10}$$

etc. (One might think that one could use a parametrization via the Euler angles, i.e.:

$$A_\Omega = \exp(i\frac{a}{\Omega} \sum_{p=1}^{\Omega} \sigma_p^{(z)}) \exp(i\frac{b}{\Omega} \sum_{p=1}^{\Omega} \sigma_p^{(y)}) \exp(i\frac{c}{\Omega} \sum_{p=1}^{\Omega} \sigma_p^{(z)}) \tag{5.11}$$

but it turns out that one cannot generate all expectation values by use of this functional).

Now $\langle A_\Omega \rangle_{B,\Omega}$ can easily be calculated:

$$\langle A_\Omega \rangle_{B,\Omega} = \Big[1 + i\Omega^{-1}\,\mathrm{th}\,\omega(a \sin\Theta \cos\phi$$

$$+ b \sin\Theta \sin\phi + c\cos\Theta + O(\Omega^{-2})\Big]^{\Omega}$$

$$\to \exp(i\,\mathrm{th}\,\omega\,\big[a \sin\Theta\cos\phi + b\sin\Theta\sin\phi$$

$$+ c\cos\Theta\big]) \tag{5.12}$$

and integration over ϕ yields

$$(2\pi)^{-1}\int_0^{2\pi} d\phi\,\langle A_\Omega\rangle_{B,\Omega} \to J_0(\mathrm{th}\,\omega\,\sin\Theta\,\sqrt{a^2+b^2})$$

$$\cdot \exp(ic\,\mathrm{th}\,\omega\cos\Theta) \tag{5.13}$$

b) The l.h.s. of (5.7)

The diagonalization of H_Ω is very simple, and inserting the matrix elements of the rotation group (which in our case are less familiar than the usual matrix elements based on the Euler angles) and paying attention to the degeneracy of the eigenvalues, one obtains

$$\mathrm{tr}\Big[e^{-H_\Omega/T} A_\Omega\Big] =$$

$$\sum_{L=0}^{\Omega/2}\sum_{L_z=-L}^{L} \frac{\Omega!\,(2L+1)}{(\frac{\Omega}{2}-L)!\,(\frac{\Omega}{2}+L+1)!}\exp\Big\{\frac{1}{T}\Big[\varepsilon(2L_z-\Omega)$$

$$+\frac{2T}{\Omega}\cdot(L(L+1) - L_z(L_z+1)\Big]\Big\}\,G_\Omega\Big(\frac{2L}{\Omega},\frac{2L_z}{\Omega};a,b,c\Big) \tag{5.14}$$

For the functions G_{Ω} we find that

$$G_{\Omega}\left(\frac{2L}{\Omega}, \frac{2L_z}{\Omega}; a,b,c\right) \left(= \langle L, L_z \,|\, A_{\Omega}(a,b,c) \,|\, L, L_z \rangle \right)$$

$$\rightarrow e^{-iwc} \, J_0\left(\sqrt{(y^2-w^2)(a^2+b^2)}\right) \left(\equiv G_{\infty}(y,w;a,b,c) \right) \tag{5.15}$$

where we introduced the intensive quantities

$$y = 2L/\Omega , \qquad w = 2L_z/\Omega \tag{5.16}$$

which are confined to the triangle in fig. 3.

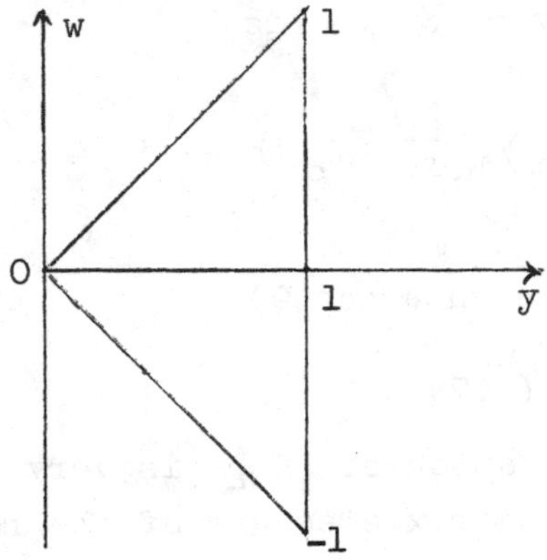

Figure 3.

We now have to calculate the limit of the double sum. The usual prescription is to replace the sum by its leading term, and indeed this can be rigorously justified in our case by considering sequences of probability measures on the triangle. What we find is that

$$\left.\begin{aligned}
&\sum_{L=0}^{\Omega/2} \sum_{L_z=-L}^{L} \rightarrow \Omega^2 \int_0^1 dy \int_{-y}^{y} dw \,, \\
&\langle A \rangle_\Omega \rightarrow \int_0^1 dy \int_{-y}^{y} dw \, \rho(y,w) \, G_\infty(y,w;\, a,b,c) \\
&\qquad \cdot \left(\int_0^1 dy \int_{-y}^{y} dw \, \rho(y,w) \right)^{-1}
\end{aligned}\right\} \quad (5.17)$$

with

$$\rho(y,w) = \exp\left(-\Omega\left[f(y_0) - f(y) + \frac{T_c}{2T}(w - w_0)^2\right]\right)$$

$$f(y) = \frac{T_c}{2T} y^2 - \frac{1-y}{2} \ln (1-y) - \frac{1+y}{2} \ln (1+y) \qquad (5.18)$$

y_0 being determined by

$$f'(y_0) = 0$$

hence

$$y_0 = \text{th} \frac{T_c}{T} y_0 \,, \quad w_0 = \frac{\varepsilon}{T_c} - \frac{1}{\Omega} \qquad (5.19)$$

and, by the usual methods,

$$\lim_{\Omega \to \infty} \langle A_\Omega \rangle_\Omega = G_\infty (y_0, w_0;\, a, b, c) \qquad (5.20)$$

The identity of the r.h.s. and the l.h.s. of (5.7) is now established if one notices the relations between ω, Θ and y_0, w_0:

$$\frac{\omega T}{T_c} = y_0 \,, \qquad w_0 = \frac{\varepsilon}{T_c} = \cos\Theta \, \text{th}\, \omega$$

$$\sin\Theta \, \text{th}\, \omega = \sqrt{\left(\frac{T\omega}{T_c}\right)^2 - \frac{\varepsilon^2}{T_c^2}} = \sqrt{y_0^2 - w_0^2} \qquad (5.21)$$

5.1 The Thermodynamic Representations

Here we shall list the various factor types which occur. The rather lengthy proofs of these statements can be found in reference 1).

(i) T = 0

$$y_0 = 1 = \text{th}\,\omega \tag{5.22}$$

The thermodynamic representation is an integral over type I factors:

$$\hat{\pi} = \int^{\Theta} d\phi \,.\, \hat{\pi}_{\phi} \tag{5.23}$$

$\hat{\pi}_{\phi}$ is a representation in an I.C.T.P.

$$\bigotimes_{p=1}^{\infty}{}^{\mathfrak{f}} \; (H_p \otimes H_p) \,, \tag{5.24}$$

all H_p being two-dimensional, such that (in an obvious notation)

$$\vec{\sigma}_p \rightsquigarrow (1 \otimes 1) \otimes \ldots . \otimes (\vec{\sigma} \otimes 1) \otimes (1 \otimes 1) . \tag{5.25}$$

p-th place (5.25)

The equivalence class is determined by

$$\otimes (\,|\, \phi > \otimes \,|\, \phi >) \tag{5.26}$$

($|\phi>$: $(\sigma^{(x)} \cos \phi + \sigma^{(y)} \sin \phi) | \phi > = | \phi >$. Phase factors are irrelevant).

That this is, in fact, the thermodynamic representation can easily be checked by comparison of the expectation values.

Since in this case

$$\otimes^{\mathfrak{s}} (H_p \otimes H_p) = \otimes^{\mathfrak{s}_1} H_p \otimes \otimes^{\mathfrak{s}_2} H_p \qquad (5.27)$$

$$(\mathfrak{s}_1, \mathfrak{s}_2 \ni \otimes \mid \phi \rangle)$$

one sees that the thermodynamic representation is just a gauge invariant integral over ampliations of the ground-state representations that we found in the foregoing lecture. (Ampliation of an algebra is defined as its tensor product with some algebra of scalars C(H).)

One remark should be added: the operator

$$\vec{s} = \lim \Omega^{-1} \sum_{p=1}^{\Omega} \vec{\sigma}_p \qquad (5.28)$$

is no longer a c-number but belongs only to the centre. (In fact, $\langle s_x \rangle = 0$ but $\langle s_x^2 \rangle \neq 0$, etc.)

(ii) $\underline{0 < T < T_0}$

The critical temperature T_0 is determined by

$$\varepsilon / T_c = \text{th } \varepsilon / T_0 \qquad (5.29)$$

(Only if $\varepsilon = 0$, $T_0 = T_c$.)

The thermodynamic representation is now an integral $\pi = \int^{\oplus} d\phi \, \pi_\phi$ over type III-factor representations in I.C.T.P.'s:

$$\otimes^{\mathfrak{s}} (H_p \otimes H_p)$$

as before, but the equivalence class $\mathcal{S}$ now contains

$$\bigotimes_{p=1}^{\infty} \sqrt{\frac{1+\text{th}\,\omega}{2}}\, |\vec{n}\rangle \otimes |\vec{n}\rangle + \sqrt{\frac{1-\text{th}\,\omega}{2}}\, |-\vec{n}\rangle \otimes |-\vec{n}\rangle$$

$$|\vec{n}\rangle : (\sigma \vec{n})\, |\pm \vec{n}\rangle = \pm\, |\pm \vec{n}\rangle$$

$$\vec{n} = \begin{bmatrix} \cos\phi \sqrt{1-(\varepsilon/T_c\ \text{th}\,\omega)^2} \\ \sin\phi \sqrt{1-(\varepsilon/T_c\ \text{th}\,\omega)^2} \\ \varepsilon/T_c\ \text{th}\,\omega \end{bmatrix} \qquad (5.30)$$

When $\|\vec{s}\| < 1!$ one gets a different type of factor since for this $\mathcal{S}$ the relation (5.27) does not hold any more (i.e. the infinite tensor product is generally not associative.

(iii) $\underline{T_0 \leq T < \infty,\ \varepsilon \neq 0}$

In this region our previous calculations have to be modified. So far we have not checked whether (y_0, w_0) is in the triangle at all. Since

$$y_0 = \text{th}\,\frac{T_c}{T}\, y_0\ , \qquad w_0 = \frac{\varepsilon}{T_c} - \frac{1}{\Omega} \quad (5.31)$$

we find that this is indeed the case as long as $T < T_0$, but not if $T \geqslant T_0$; see fig. 4.
We have $|w_0| \geqslant y_0$ in the latter case. The maximum value inside the triangle is attained on

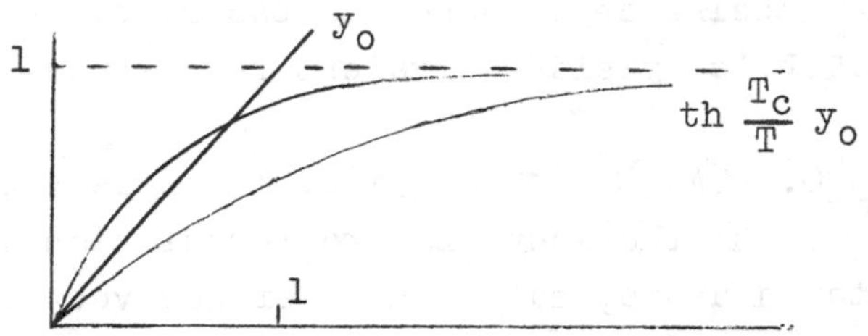

Figure 4.

the boundary $|w| = y$, and therefore

$$y_o = \text{th}\ \varepsilon/T\ , \quad \sin\Theta = 0$$

$$\langle A_\Omega \rangle_\Omega \to e^{ic\ \text{th}\ \varepsilon/T} \qquad (5.32)$$

One sees that there is a phase transition at T_o. The thermodynamic representation is again in an I.C.T.P.

$$\bigotimes_{p=1}^{\infty}{}^{\xi} (H_p \otimes H_p)$$

but it is now no longer integrated, $(\sin\Theta = 0)$, but a type III-factor representation. ξ is determined by

$$\bigotimes_{p=1}^{\infty} \left(\sqrt{\frac{1+\ \text{th}\ \varepsilon/T}{2}}\ |\vec{z}\rangle \otimes |\vec{z}\rangle + \sqrt{\frac{1-\ \text{th}\ \varepsilon/T}{2}}\ |-\vec{z}\rangle \otimes |-\vec{z}\rangle \right)$$

$$(|\pm\vec{z}\rangle : \sigma^{(z)}\ |\pm\vec{z}\rangle = \pm|\vec{z}\rangle)\ . \qquad (5.33)$$

(One could equally use the reference vector

$$\bigotimes_{p=1}^{\infty} \sqrt{\frac{1+\text{th}\ \varepsilon/T}{2}}\ |\vec{z}\rangle \otimes |e_1\rangle + \sqrt{\frac{1-\text{th}\ \varepsilon/T}{2}}\ |-\vec{z}\rangle \otimes |e_2\rangle)$$

$\{|e_1\rangle, |e_2\rangle\}$ being an orthonormal basis, since all these I.C.T.P.'s yield equivalent representations.)

When $\underline{\varepsilon = 0}$, $\langle A_\Omega \rangle_\Omega \to 1$, i.e. there is complete disorder. The thermodynamic representation is of the type stated above, but with reference vector

$$\bigotimes_{p=1}^{\infty} \left(\frac{1}{\sqrt{2}} \quad |\vec{z}\rangle \otimes |\vec{z}\rangle \quad + \frac{1}{\sqrt{2}} \quad |-\vec{z}\rangle \otimes |-\vec{z}\rangle \right) \quad (5.34)$$

i.e. a type II_1 factor representation. Here we have the remarkable situation that beyond T_c the hamiltonian vanishes in some sense and a II_1 representation becomes possible.

(iv) $\underline{T = \infty}$

We obtain the same type II_1 factor representation as above in both cases $\varepsilon = 0$ and $\varepsilon \neq 0$.

5.2 An "Isotropic" BCS-Model

One can apply the above-developed techniques to the following hamiltonian.

$$H_\Omega = \varepsilon \sum_{p=1}^{n} \sigma_p^{(z)} - \frac{T_c}{2\Omega} \sum_{p,p'=1}^{\Omega} \vec{\sigma}_p \vec{\sigma}_{p'}, \; (\varepsilon > 0) \quad (5.35)$$

We arrive at the equality

$$\operatorname{th} \omega = {}^{T}/_{T_c} \; \omega - {}^{\varepsilon}/_{T_c} \quad (5.36)$$

or $(y_0 = \operatorname{th} \omega)$

$$y_0 = \operatorname{th} \omega \left({}^{T_c}/_{T} \; y_0 + {}^{\varepsilon}/_{T} \right) \quad (5.37)$$

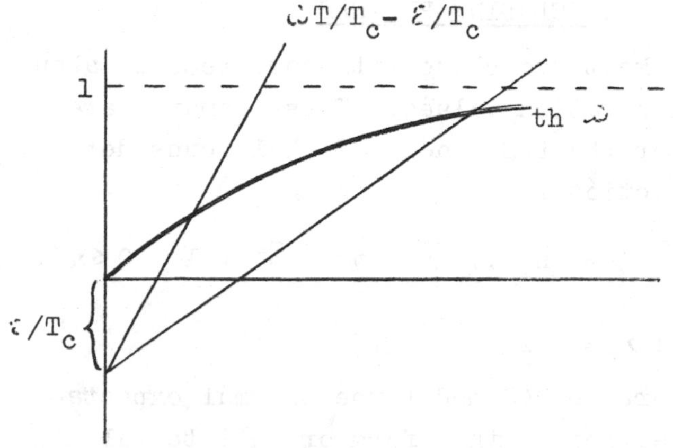

Figure 5.

and $\lim \langle A_\Omega \rangle_\Omega \rightarrow e^{iy_0 c}$.

The thermodynamic representation is a

type I factor representation for $T = 0$

type III factor representation for $0 < T < \infty$

type II_1 factor representation for $T = \infty$.

There is no phase transition.

If, however, $\varepsilon = 0$, then the calculation yields

$$\langle A_\Omega \rangle_\Omega \rightarrow \frac{\sin y_0 \sqrt{a^2 + b^2 + c^2}}{y_0 \sqrt{a^2 + b^2 + c^2}}$$

$$= \lim \frac{1}{4\pi} \int \sin \theta \, d\theta \, d\phi \, \langle A_\Omega \rangle_B \tag{5.38}$$

hence the thermodynamic representation is an integral of type III (type I) factor representations if $0 < T < T_0$ $(T = 0)$. At T_0 there is a phase transition, and for $T \geqslant T_0$ $\langle A_\Omega \rangle_\Omega \rightarrow 1$ hence we have a type II_1 factor representation.

6. CONCLUDING REMARKS

I would like to conclude with some remarks about the thermal expectation values. These remarks are not specific for the BCS model. I shall consider the general functional

$$\langle \vec{\sigma}_1 \ldots \vec{\sigma}_j \rangle = \vec{n}_1 \ldots \vec{n}_j \, \eta^j \qquad \vec{n}^2 = 1 \qquad 0 \leq \eta \leq 1$$

$$\langle 1 \rangle = 1$$

We know that for the BCS model the thermal expectation value is either of this form or an integral over it.

(i) The functional is clearly bounded and can therefore be uniquely extended to a bounded linear functional over the norm completion of the *-algebra generated by the $\vec{\sigma}$'s. This completed algebra is a C^*-algebra.

(ii) In a particular representation we may form the weak closure of the (representation of the) C^*-algebra. According to the Hahn-Banach theorem, the functional can be extended to a norm-continuous functional over the whole von Neumann algebra. (One can even suppose this extension to be positive, which is a theorem of Krein.) Whether this extension will be normal or not depends on the representation. In the one obtained by the GNS construction, it is of course even weakly continuous since it is a diagonal matrix element. On the other hand, in the irreducible I.C.T.P. representation with "spins up", it is not weakly continuous. From

$$P_N = \prod_{i=N}^{\infty} \frac{1+\sigma_i^{(z)}}{2} \tag{5.39}$$

This product converges strongly towards the projector on the subspace where for $i \geq N$ all spins are up. $\underset{N\to\infty}{\text{s-lim}}\ P_N = 1$, on the other hand

$$\langle P_N \rangle = \lim_{N'\to\infty} \prod_{i=N}^{N'} \left\langle \frac{1+\sigma_i^{(z)}}{2} \right\rangle = \lim_{N'\to\infty} \left(\frac{1+\gamma}{2}\right)^{N'-N} = 0$$

$$\text{if } \gamma < 1. \tag{5.40}$$

$$\Rightarrow \lim_{N\to\infty} \langle P_N \rangle \neq \langle \lim_{N\to\infty} P_N \rangle$$

(iii) One may raise the question of the existence of a density matrix. We have noted for the ground state that $\lim_{\Omega\to\infty} H_{B,\Omega}$ exists strongly and the same holds in the thermal representation, provided a suitable element from the commutant is subtracted. Thus $\lim_{\Omega\to\infty} e^{-\beta H_{B,\Omega}}$ exists but is not in the trace class, and consequently,

$$\lim_{\Omega\to\infty} e^{-\beta H_{B,\Omega}} / \operatorname{tr} \lim_{\Omega\to\infty} e^{-\beta H_{B,\Omega}} = 0 \tag{5.41}$$

(iv) Whether the above-constructed functional can be written as the trace of a density matrix times the element of the algebra depends on the representation. In the thermal representation we have $\langle A \rangle = \operatorname{tr} P_0 A$, P_0 being the projector onto the cyclic vector. However, generally $P_0 \notin$ the von Neumann algebra since we are in a type III factor which does not contain finite-dimensional projectors.

In the representations where the extensions of the functional is not normal, it certainly cannot be written as $\langle A \rangle = \operatorname{tr} \rho A$, since this would imply normality.

(v) An exception is the case $\eta = 0$, e.g. $\langle \vec{\sigma}_i \rangle = 0$. This functional is already a trace, and the thermal representation is a II_1 factor.

(vi) Although the expectation value cannot be written $\langle A \rangle = \operatorname{tr}(e^{-\beta H} A) / \operatorname{tr}(e^{-\beta H})$, the KMS analyticity property is satisfied in our model.

REFERENCES

Infinite Tensor Products

1) J. von Neumann, "On infinite direct products", Comp. Math. 6, 1 (1938).

2) A collection of the main results can be found in M. Guenin, A. Wehrl, W. Thirring, "Introduction to Algebraic Techniques," CERN, 1968-69.

Traces

3) J. Dixmier, "Des algebres d'opérateurs dans l'espace hilbertien", (Gauthier Villars, Paris, 1957).

The BCS-Model

Most standard texts on super-conductivity. The calculations mentioned here can be found in more detail in

4) W. Thirring and A. Wehrl, Commun. Math. Phys. 4, 303 (1967), for T = 0.

5) W. Thirring, Commun. Math. Phys. 7, 181 (1968) for T ≠ 0; this paper contains also a list of further references.

6) F. Jelinek, Commun. Math. Phys. 9, 169 (1968).

7) Some unpublished results are from F. Jelinek and A. Wehrl.

FUNCTIONAL INTEGRATION METHODS IN QUANTUM MECHANICS

IZURU FUJIWARA

Institutt for Teoretisk Fysikk
N.T.H. TRONDHEIM

INTRODUCTION

For functional integration methods in quantum mechanics there is a nice text book of Feynman and Hibbs [1], in which Chapters 10-12 give introductions to statistical mechanics, variational method and other problems in probability. It is also very useful for people who are interested in probability theory (and its application to physical sciences) to refer to Kac's book [2]. Since Feynman's heuristic proposal of his path-integrals in non-relativistic quantum mechanics [3] a number of papers have appeared treating related topics, some of which are included in reference [4]. Unfortunately, however, the study of the subject has not yet been developed in a completely satisfactory way and there are still many things left to be done.

The first aim of the present lectures is to reinvestigate Feynman's idea of the probability amplitude of a space-time path from an observational point of view. Some interesting features of such a probability amplitude will be discussed. Then with the aid of an exact definition of the path-integral we shall look at the relationship between classical mechanics and quantum mechanics, finally showing how to perform functional integrations in quantum mechanics on the basis of the author's works in 1963 [5],(parts of this were made in collaboration with Prof. P.Chr. Hemmer [6]). These results are contained in none of the works cited above.

It is the hope of the author that the present lectures will serve as a brief introduction to the subject and help you to appreciate beauty and elegance in the functional integral formulation of quantum mechanics, though it might seem at first sight, to be fairly complicated and ugly.

1. VIRTUAL AMPLITUDES OF SUCCESSIVE TRANSITIONS

In his famous paper "Space-Time Approach to Non-Relativistic Quantum Mechanics" [3], Feynman proposed the concept of a probability amplitude associated with a completely specified motion as a function of time. To explain this idea we shall first consider an experiment in which we can make three measurements successively in time: first of a quantity A, then of B, and then of C. For example these might be three successive position measurements. Suppose that a, b and c are respectively one of the possible results of A, B and C and define P_{ab} as the probability that if measurement A gave the result a, the measurement B will give the result b. Similarly, P_{bc} is the probability that if measurement B gives the result b, then measurement C gives c. Further, let P_{ac} be the chance that if A gives a, then C gives c. Finally, denote by P_{abc} the probability of all three, i.e., if A gives a, then B gives b, and C gives c. If the events between a and b are independent of those between b and c, then

$$P_{abc} = P_{ab}P_{bc}. \tag{1.1}$$

and we expect that

$$P_{ac} = \sum_b P_{abc}, \tag{1.2}$$

where we sum, or integrate, over all the mutually exclusive alternatives for b. This is because, if, initially, measurement A gives a and the system is later found to give the result c to measurement C, the quantity B must have had some value, say one of the b's, at the time intermediate to A and C. The probability that it was b is P_{abc}. In classical physics it is always true, but in quantum mechanics it is often false.

In quantum mechanics there exist complex numbers ϕ_{ab} and ϕ_{bc} such that

$$P_{ab} = |\phi_{ab}|^2 \text{ and } P_{bc} = |\phi_{bc}|^2 \tag{1.3}$$

These numbers are obtained as inner product of eigenvectors

$$\phi_{ab} = \langle b|a\rangle \text{ and } \phi_{bc} = \langle c|b\rangle , \tag{1.4}$$

where the eigenvectors are defined by

$$A|a\rangle = a|a\rangle , \; B|b\rangle = b|b\rangle \text{ and } C|c\rangle = c|c\rangle . \tag{1.5}$$

If the ket $|b\rangle$ represents a non-degenerate state, we have

$$\sum_b |b\rangle\langle b| = 1 \text{ (unit operator)}, \tag{1.6}$$

and accordingly

$$\phi_{ac} = \langle c|a\rangle = \sum_b \langle c|b\rangle\langle b|a\rangle = \sum_b \phi_{ab}\phi_{bc}. \tag{1.7}$$

At this point we have to distinguish three different alternatives from the point of view of measuring the quantity B.

(i) If B is measured between the experiments A and C to yield one of its possible results, say b, or more precisely if there exists a one-to-one correspondence between the state $|b\rangle$ of the system under consideration and a state $|\beta\rangle$ of a second apparatus system which can interact with the first, then, after detecting the state $|\beta\rangle$, we can extract one particular projection operator $|b\rangle\langle b|$ out of the unit operator (1.6), and accordingly equation (1.7) becomes

$$\phi_{a(\beta)c} = \phi_{ab}\phi_{bc}. \tag{1.8}$$

This possibility of the correspondence between $|\beta\rangle$ and $|b\rangle$ is usually called the projection postulate. In this ideal case we will have relationships essentially the same as the classical ones (1.1) and (1.2), that is

$$P_{a(\beta)c} = |\phi_{a(\beta)c}|^2 \text{ and } P_{ac} = \sum_\beta P_{a(\beta)c}. \tag{1.9}$$

The summation over all possible values of β is equivalent to summing over all possible values of b.

(ii) The second case may be quite usual in quantum mechanics, where more than one object state $|b\rangle$ will correspond to a particular state $|\beta\rangle$ of the second apparatus system. In other words, after interaction with the apparatus system the quantum-mechanical state of the object system is represented by a superposition of eigenvectors $|b\rangle$ over a certain range of b. Then we have from (1.7)

$$\phi_{ac} = \sum_{(b)} \phi_{ab}\,\phi_{bc} \;,$$

where the symbol (b) denotes a restricted range of summation or integration. In this case we can not have any well-defined joint amplitude ϕ_{abc}. This is the most interesting case from the point of view of quantum-mechanical measurement. We shall, however, not enter into more details of this problem at the moment, dealing only with the functional integration methods.

(iii) If we do not measure B at all, or if the measurement of B is simply unsuccessful in the sense that the inspection of $|\beta\rangle$ fails to restrict the range of summation or integration in (1.7), then we have the case which Feynman investigated in his 1948 paper. Even in this case we have an amplitude

$$\Phi_{abc} = \phi_{ab}\phi_{bc} \tag{1.10}$$

such that

$$\phi_{ac} = \sum_b \Phi_{abc} , \tag{1.11}$$

and we may be allowed to introduce a joint probability such as

$$P^V{}_{abc} = |\Phi_{abc}|^2. \tag{1.12}$$

But in general we shall have the rather unfortunate result that

$$P_{ac} = |\phi_{ac}|^2 \neq \sum_b P^V{}_{abc} ,$$

in contrast to the <u>actual</u> amplitude $\phi_{a(\beta)c}$, satisfying (1.9). The number Φ_{abc} must be interpreted as a <u>virtual</u> amplitude, since the states $|b\rangle$ appearing in the decomposition (1.7) are all virtual states. In this case we can not assume that to get from a to c the system had to go through a condition such that B had some definite value, b, simply because the statement, "B had some value", is meaningless. In deriving (1.9) it must be certain that in each experiment giving a and c, B had some value.

The composition law (1.7) is a typical representation of the wave nature of matter or of the Huygens principle in a generalized sense. The quantum-mechanical formalism is the only way of describing correctly the interference phenemena displayed by matter wave, and just for this purpose the virtual amplitudes such as Φ_{abc} play an essential part.

The virtual amplitude Φ_{abc} can be normalized when it is divided by the square root of

$$P^V{}_{ac} = \sum P^V{}_{abc} \neq P_{ac}.$$

We thus have

$$1 = \sum_b \left| \Phi_{abc} \Big/ \sqrt{P^V{}_{ac}} \right|^2 \qquad (1.13)$$

In the following we shall investigate peculiar properties of Φ_{abc} by taking as an example a free transition of a non-relativistic particle of mass m in one dimension. Let $|\psi\rangle$ denote the Heisenberg wave packet with minimum uncertainty product, $\Delta p \Delta q = \hbar/2$. Its momentum and position representations read respectively

$$\langle p|\psi\rangle = (2\pi\chi^2)^{-\frac{1}{4}} \exp\left[-(p-p_0)^2/4\chi^2 - ipq_0/\hbar\right] \qquad (1.14)$$

and

$$\langle q|\psi\rangle = (2\pi\lambda^2)^{-\frac{1}{4}} \exp\left[-(q-q_0)^2/4\lambda^2 + ip_0(q-q_0)/\hbar\right] \qquad (1.15)$$

with $\lambda\chi = \hbar/2$. The expectation values are $\langle p\rangle = p_0$ and $\langle q\rangle = q_0$. The standard deviations are $\Delta p = \chi$ and $\Delta q = \lambda$, as is well known. In the Heisenberg representation we have

$$\langle p,t|\psi\rangle = \langle p|e^{-itH/\hbar}|\psi\rangle = e^{-itp^2/2m\hbar}\langle p|\psi\rangle \tag{1.16}$$

and

$$\langle q,t|\psi\rangle = \int dp\, \langle q,t|p,t\rangle\langle p,t|\psi\rangle \tag{1.17}$$

$$= (2\pi\lambda^2)^{-\frac{1}{4}}(1+it/T)^{-\frac{1}{2}}\exp\left[-\frac{(q-q_0-tp_0/m)^2}{4\lambda^2\{1+(t/T)^2\}}\right]$$

$$\exp\left\{\frac{1}{1+(t/T)^2}\left[\frac{i}{\hbar}\,p_0(q-q_0) - \frac{it}{2m\hbar}\,p_0{}^2 + \frac{it}{4\lambda^2 T}(q-q_0)^2\right]\right\}$$

with $T = m\hbar/2\chi^2$.

Now we shall assume that the initial state $|\psi_0\rangle$ is the above $|\psi\rangle$ with $q_0=0$ and that the final state $|\psi_1\rangle$ is the $|\psi\rangle$ wherein $q_0 = \tau p_0/m$. Here τ denotes the classical arrival time. Then we can think of two kinds of virtual amplitudes, namely,

$$\Phi(p) = \langle\psi_1,\tau|p,t\rangle\langle p,t|\psi_0,0\rangle \tag{1.18}$$

$$= \langle\psi_1,0|p,t-\tau\rangle\langle p,t|\psi_0,0\rangle$$

and

$$\Phi(q) = \langle\psi_1,\tau|q,t\rangle\langle q,t|\psi_0,0\rangle \tag{1.19}$$

$$= \langle\psi_1,0|q,t-\tau\rangle\langle q,t|\psi_0,0\rangle .$$

We then have

$$|\Phi(p)|^2 = |\langle p|\psi_1\rangle|^2\,|\langle p|\psi_0\rangle|^2 \tag{1.20}$$

$$= (2\pi\chi^2)^{-1}\exp\left[-(p-p_0)^2/\chi^2\right]$$

and

$$|\Phi(q)|^2=(2\pi\lambda^2)^{-1}\left[1+(t/T)^2\right]^{-\frac{1}{2}}\left[1+(t^1/T)^2\right]^{-\frac{1}{2}} \qquad (1.21)$$

$$\exp\left[-\frac{(q-tp_0/m)^2}{2\lambda^2\{1+(t/T)^2\}}\right]\cdot\exp\left[-\frac{(q-tp_0/m)^2}{2\lambda^2\{1+(t^1/T)^2\}}\right]$$

with $t^1=\tau-t$. If we assume $\tau/T \ll 1$, then (1.21) can be approximated by

$$|\Phi(q)|^2 = (2\pi\lambda^2)^{-1}\exp\left[-(q-tp_0/m)^2/\lambda^2\right]. \qquad (1.22)$$

Using the integral formulas

$$\int_{-\infty}^{\infty} dx\; e^{-x^2/2\sigma^2} = \sqrt{2\pi\sigma},\quad \int_{-\infty}^{\infty} dx\; xe^{-x^2/2\sigma^2} = 0 \text{ and}$$

$$\int_{-\infty}^{\infty} dx\; x^2\; e^{-x^2/2\sigma^2} = \sqrt{2\pi\sigma^3}$$

one easily finds that

$$\langle p(t)\rangle = \int dp\; p|\Phi(p)|^2 \Big/ \int dp|\Phi(p)|^2 = p_0,$$

$$[\Delta p(t)]^2 = \int dp(p-\langle p(t)\rangle)^2|\Phi(p)|^2 \Big/ \int dp|\Phi(p)|^2 = \frac{\chi^2}{2},$$

$$\langle q(t)\rangle = \int dq\; q|\Phi(q)|^2 \Big/ \int dq|\Phi(q)|^2 = \frac{t}{m}p_0,$$

$$[\Delta q(t)]^2 = \int dq(q-\langle q(t)\rangle)^2|\Phi(q)|^2 \Big/ \int dq|\Phi(q)|^2 = \frac{\lambda^2}{2},$$

and accordingly

$$\Delta p(t)\Delta q(t) = \lambda\chi/2 = \hbar/4 < h/2 \qquad (1.23)$$

This result is not surprising at all, since our virtual amplitudes never refer to the outcomes of actual measurements. Heisenberg's uncertainty relationship $\Delta p \Delta q \geqslant \hbar/2$ is established on the basis of the state functions $\psi(p) = \langle p|\psi\rangle$ and $\psi(q) = \langle q|\psi\rangle$, which are related to each other by the Fourier analysis:

$$\psi(q) = \int dp \langle q|p\rangle \langle p|\psi\rangle.$$
$$= h^{-\frac{1}{2}} \int dp \exp (ipq/\hbar) \psi(p).$$

In contrast to the actual amplitudes $\psi(p)$ and $\psi(q)$ our virtual amplitudes $\Phi(p)$ and $\Phi(q)$ do not satisfy such a relationship, and needless to say the standard deviations that they yield cannot fulfil Heisenberg's requirement.

2. CLASSICAL LIMIT VIA VIRTUAL AMPLITUDES

The virtual amplitudes introduced in the last part are the most convenient for discussing the classical limit of the quantum-mechanical formalism. It is interesting to see how it reduces to that of classical mechanics in the formal limit $h \to 0$. The classical limit of virtual amplitudes was discussed first by Dirac [7] in 1933, and his remarks were the starting point of Feynman's path-integral formulation of quantum mechanics. Since the merit of this formulation lies in enabling us to see directly the interrelationship of quantum mechanics to the Lagrangian formalism of classical mechanics, we shall first try to look at it in a simpler version.

The amplitude for the transition of a microscopic system from an initial state ψ at time $t = t_0$ to a final state $\emptyset$ at a later time $t = t_1$ is given by the inner product of state vectors $|\psi,t_0\rangle$ and $|\emptyset,t_1\rangle$, which are defined in the Heisenberg representation by

$$|\psi,t_o\rangle = \int dq_o\, \psi(q_o)|q_o,t_o\rangle$$

and

$$|\phi,t_1\rangle = \int dq_1\, \phi(q_1)|q_1,t_1\rangle\,.$$

Here $|q',t\rangle$ denotes an eigenvector of the Heisenberg operator $q(t)$ such that $q(t)|q',t\rangle = q'|q',t\rangle$, and we have restricted ourselves to the case of one-dimensional motion. In terms of the kernel

$$K(q_1,t_1;q_o,t_o) \equiv \langle q_1,t_1|q_o,t_o\rangle \qquad (2.1)$$

the above transition amplitude can be rewritten as

$$\begin{aligned}&\langle\phi,t_1|\psi,t_o\rangle\\ &=\int dq_1\int dq_o\langle\phi,t_1|q_1,t_1\rangle\langle q_1,t_1|q_o,t_o\rangle\langle q_o,t_o|\psi,t_o\rangle\\ &=\int dq_1\int dq_o\ \phi^*(q_1)\ K(q_1,t_1;q_o,t_o)\psi(q_o)\ ,\end{aligned}$$

where use is made of the relationship $\int dq'|q',t\rangle\langle q',t| = 1$, (the unit operator).

As is well known, the kernel K is a solution to the Schrödinger equation

$$\left[i\hbar\frac{\partial}{\partial t_1} - H(p_1,q_1)\right]K(q_1,t_1;q_o,t_o) = 0 \qquad (2.2)$$

(with $p_1 = -i\hbar\partial/\partial q_1$), satisfying the initial condition

$$\lim_{t_1\to t_o} K(q_1,t_1;q_o,t_o) = \langle q_1,t_o|q_o,t_o\rangle = \delta(q_1-q_o).\ (2.3)$$

If the Hamiltonian operator does not depend on time explicitly, the kernel is written down at once as

$$K(q_1,t_1;q_0,t_0) = e^{-i\tau H(p_1,q_1)/\hbar}\delta(q_1 - q_0) \qquad (2.4)$$

with $\tau = t_1 - t_0$. For the case of a free particle it is quite simple to evaluate (2.4) to obtain

$$K_f(q_1,t_1;q_0,t_0) = \left(\frac{m}{ih\tau}\right)^{\frac{1}{2}} \exp\left[\frac{im}{2\hbar\tau}(q_1-q_0)^2\right], \qquad (2.5)$$

where we have made use of a very useful formula:

$$\begin{aligned} e^{\frac{a}{2}\left(\frac{\partial}{\partial x}\right)^2}\delta(x) &= e^{\frac{a}{2}\left(\frac{\partial}{\partial x}\right)^2}\int_{-\infty}^{\infty} du\, e^{2\pi iux} \\ &= e^{-\frac{1}{2a}x^2}\int_{-\infty}^{\infty} du\, e^{-2\pi^2 a(u-ix/2\pi a)^2} \\ &= \frac{1}{\sqrt{2\pi a}}\, e^{\left[-\frac{1}{2a}x^2\right]} . \end{aligned} \qquad (2.6)$$

In (2.5) we have to set $a = i\hbar\tau/m$. In the case of a harmonic oscillator the Hamiltonian is $H = \frac{1}{2m}p^2 + \frac{m}{2}\omega^2 q^2$, and we can have a nice factorization of the evolution operator [8]:

$$\begin{aligned} e^{-itH/\hbar} &= e^{\frac{i\hbar\tau}{2m}\left(\frac{\partial}{\partial q}\right)^2 - \frac{im\tau}{2\hbar}\omega^2 q^2} \\ &= e^{f(\tau)q^2}\, e^{g(\tau)(\partial/\partial q)^2}\, e^{f(\tau)q^2} \end{aligned} \qquad (2.7)$$

with $f(\tau) = \frac{m\omega}{2i\hbar}\tan(\omega\tau/2)$ and $g(\tau) = \frac{i\hbar}{2m\omega}\sin(\omega\tau)$. This is easily obtained by differentiating both sides of (2.7) to get differential equations for the functions $f(\tau)$ and $g(\tau)$, which must satisfy the initial conditions $f(0)=g(0)=0$. Then (2.6) will yield the result that the kernel (2.4) is equal to

$$K_c(q_1,t_1;q_0,t_0) = \left[\frac{i}{h}\frac{\partial^2 S}{\partial q_1 \partial q_0}\right]^{\frac{1}{2}} e^{\frac{i}{h}S} \qquad (2.8)$$

where S denotes the classical action function. This is the classical Lagrangian function $L(\dot{q},q)$ integrated with respect to time along the actual classical path $q_c(t)$ connecting the end points (q_1,t_1) and (q_0,t_0):

$$S(q_1,t_1;q_0,t_0) = \int_{t_0}^{t_1} dt\, L[\dot{q}_c(t),q_c(t)] \,. \qquad (2.9)$$

In the case of the harmonic oscillator with the Lagrangian function $L = \frac{m}{2}\dot{q}^2 - \frac{m}{2}\omega^2 q^2$ the classical path

$$q_c(t) = \frac{1}{\sin\omega\tau}\left[q_1 \sin\omega(t-t_0) + q_0 \sin\omega(t_1-t)\right]$$

gives the classical action

$$S(q_1,t_1;q_0,t_0) = \frac{m\omega}{2\sin\omega\tau}\left[(q_1{}^2+q_0{}^2)\cos\omega\tau - 2q_1q_0\right],$$

where $\tau = t_1 - t_0$. If we take the limit $\omega \to 0$, this will reduce to the case of a free particle, and $K=K_c$ is valid for this case too. The quantity K_c, which is written down simply in terms of the classical action, is conveniently termed the semi-classical kernel. If $d^3V(q)/dq^3=0$, we have the result $K=K_c$ without any correction terms. It is one of the tasks of our functional integration method to prove this assertion.

Now we shall investigate the properties of the virtual amplitudes

$$\Phi(p) = \langle q_1,t_1|p,t\rangle \langle p,t|q_0,t_0\rangle \qquad (2.10)$$

and

$$\Phi(q) = \langle q_1,t_1|q,t\rangle \langle q,t|q_0,t_0\rangle \qquad (2.11)$$

for the case of a free particle. Both are complex quantities and satisfy the relationships

$$\int dp\, \Phi(p) = \int dq\, \phi(q) = \langle q_1, t_1 | q_0, t_0 \rangle = K_f(q_1, t_1; q_0 t_0) \tag{2.12}$$

when $K_f(q_1, t_1; q_0 t_0)$ is given in (2.5)

Since

$$\langle p, t | q_0, t_0 \rangle = \langle p | e^{-i(t-t_0)H/\hbar} | q_0 \rangle$$

$$= \exp\left[-\frac{i(t-t_0)p^2}{2m\hbar} \right] \langle p | q_0 \rangle$$

$$= \frac{1}{\sqrt{h}} \exp\left[-\frac{i(t-t_0)p^2}{2m\hbar} - \frac{i}{\hbar}\, pq_0 \right]$$

and

$$\langle q_1, t_1 | p, t \rangle = \frac{1}{\sqrt{h}} \exp\left[\frac{i(t-t_1)p^2}{2m\hbar} + \frac{i}{\hbar}\, pq_1 \right],$$

we have

$$\Phi(p) = \frac{1}{h} \exp\left[-\frac{i\tau p^2}{2m\hbar} + \frac{i}{\hbar}\, p\, (q_1 - q_0) \right]$$

$$= \frac{1}{h} \exp\left[\frac{im}{2\hbar\tau} (q_1 - q_0)^2 \right] \exp\left[-\frac{i\tau}{2m\hbar} \left\{ p - \frac{m}{\tau} (q_1 - q_0) \right\}^2 \right]$$

and accordingly, using (2.6)with $a = m\hbar/i\tau$,

$$\frac{\Phi(p)}{K_f(q_1, t_1; q_0, t_0)} = \left(\frac{i\tau}{mh}\right)^{\frac{1}{2}} \exp\left[-\frac{i\tau}{2m\hbar} \left\{ p - \frac{m}{\tau}(q_1 - q_0) \right\}^2 \right]$$

$$= \exp\left[\frac{m\hbar}{2i\tau} \left(\frac{\partial}{\partial p}\right)^2 \right] \delta\left[p - \frac{m}{\tau} (q_1 - q_0) \right] \tag{2.13}$$

In the limit $h \to 0$ the righthand side tends to $\delta\,[p - m(q_1 - q_0)/\tau]$, which means that the Huygens principle (2.12) reduces to the particle picture of of classical determinism.

If we use the result (2.5) exactly the same thing can be done for the virtual amplitude (2.11). Because we have

$$\frac{(q_1-q)^2}{t_1-t} + \frac{(q-q_o)^2}{t-t_o} = \frac{(q_i-q_o)^2}{\tau} + \frac{\tau(q-q_c)^2}{(t_1-t)(t-t_o)} \tag{2.14}$$

with the classical path

$$q_c(t) = \frac{1}{\tau}\left[(t-t_o)q_1 + (t_1-t)q_o\right] ,$$

we shall at once obtain

$$\frac{\Phi(q)}{K_f(q_1,t_1;q_o,t_o)} = \left[\frac{m\tau}{ih(t_1-t)(t-t_o)}\right]^{\frac{1}{2}}$$

$$\cdot \exp\left[\frac{im}{2\hbar}\frac{\tau}{(t_1-t)(t-t_o)}\left\{q-q_c(t)\right\}^2\right] \tag{2.15}$$

$$= \exp\left[\frac{i\hbar(t_1-t)(t-t_o)}{2m\tau}\left(\frac{\partial}{\partial q}\right)^2\right]\delta\left[q-q_c(t)\right]$$

again using (2.6) with $a=i\hbar(t_1-t)(t-t_o)/m\tau$.

In the formal limit $h \to 0$ the righthand side reduces again to a delta function, which demands that the intermediate point must be the classical point $q_c(t)$.

These considerations explain the origin of the principle of stationary action in classical mechanics. As is well known, the variational principle

$$\delta\int_{t_o}^{t_1} dt\; L[\dot{q}(t),q(t)] = 0 \tag{2.16}$$

leads to the Lagrangian equation of motion determining

the actual classical path $q_c(t)$ passing through a pair of end points (q_0,t_0) and (q_1,t_1). In (2.14) we have assumed a fictitious motion of a free particle travelling first from (q_0,t_0) to an arbitrary intermediate point (q,t) and then from (q,t) to (q_1,t_1). Equation (2.14) is proportional to the action integral evaluated for the totality of this fictitious motion. The actual intermediate point q for a given value of t is fixed by the condition that the action must be stationary for small variations in q. The righthand side of (2.14) is differentiated with respect to q and then equated to zero leads at once to $q=q_c(t)$. If we assume an infinite sequence of intermediate points between (q_0,t_0) and (q_1,t_1), then a straightforward generalization of the above proccedures [9] will lead to the variational principle, which is formulated in the form of equation (2.16).

3. VIRTUAL AMPLITUDE FOR SPACE-TIME PATH

In the last part we investigated virtual amplitudes for the case where the total time interval t_1-t_0 is subdivided into two parts $t-t_0$ and t_1-t. Now we rewrite (2.12) as

$$K(q_1,t_1;q_0,t_0) \equiv \langle q_1,t_1|q_0,t_0\rangle = \int dq \langle q_1,t_1|q,t\rangle \langle q,t|q_0,t_0\rangle = \int dq\, K(q_1,t_1;q,t)K(q,t;q_0,t_0). \tag{3.1}$$

This composition law is quite universal being valid for any dynamical system. Let a total time interval t_b-t_a be subdivided into n equal parts τ : $t_b-t_a= n\tau$, and set $t_k=t_a+k\tau$ with $t_a=t_0$ and $t_b=t_n$. Then repeated use of the composition rule (3.1) gives an (n-1)-fold integral

$$K(q_b,t_b;q_a,t_a) = \underbrace{\int\cdots\int}_{(n-1)} \prod_{j=1}^{n-1} dq_j \prod_{k=o}^{n-1} K(q_{k+1},t_{k+1};q_k,t_k) \tag{3.2}$$

with $q_0=q_a$ and $q_n=q_b$. If the number n is increased without limit for a fixed value of t_b-t_a, then τ becomes infinitesimally small and each kernel in (3.2) is approximately, to first order in τ ,

$$K(q_{k+1},t_{k+1};q_k,t_k)=e^{-\frac{i\tau}{\hbar}V(q_{k+1})}\, e^{\frac{i\hbar\tau}{2m}(\partial/\partial q_{k+1})^2}\,\delta(q_{k+1}-q_k) = \sqrt{\frac{m}{ih\tau}}\; e^{\frac{i\tau}{\hbar}\left[\frac{m}{2}\left\{(q_{k+1}-q_k)/\tau\right\}^2 - V(q_{k+1})\right]} \tag{3.3}$$

where use is again made of (2.6) and the Hamiltonian is the ordinary nonrelativistic one: $H=\frac{1}{2m}p^2 + V(q)$. In the limit of infinite subdivision ($n\to\infty$) the sequence of the space-time points (q_k,t_k) goes over to a fictitious path $q(t)$ with $t_b \geq t \geq t_a$ and such that $q(t_a)=q_a$ and $q(t_b)=q_b$. The virtual amplitude for a path $q(t)$ is written as

$$\Phi[q(t)] = \lim_{n\to\infty} \prod_{k=0}^{n-1} K(q_{k+1},t_{k+1};q_k,t_k), \qquad (3.4)$$

which is a functional in $q(t)$. If we write

$$\lim_{n\to\infty} \underbrace{\int\cdots\int}_{(n-1)} \prod_{j=1}^{n-1} dq_j = \int_F d[q(t)] , \qquad (3.5)$$

we have from (3.2) a functional integral representation of the kernel

$$K(q_b,t_b;q_a,t_a) = \int_F d[q(t)]\, \Phi[q(t)] , \qquad (3.6)$$

which is the sum over all possible paths according to Feynman. From (3.3) we see that the functional integrand $\Phi[q(t)]$ is given by

$$\Phi[q(t)] = N(t_b,t_a)\exp\left[\frac{i}{\hbar}\int_{t_a}^{t_b} dt\; L[\dot{q}(t),q(t)]\right] \qquad (3.7)$$

with the classical Lagrangian function

$$L(\dot{q},q) = \frac{m}{2}\dot{q}^2-V(q) \qquad (3.8)$$

and an infinite normalization constant

$$N(t_b,t_a) = \lim_{n\to\infty} \left(\frac{m}{ih\tau}\right)^{n/2} . \qquad (3.9)$$

With the aid of the virtual amplitude $\Phi[q(t)]$ the Hamiltonian formalism has thus been transformed into a Lagrangian formalism.

At first sight the appearance of an infinite normalization factor $N(t_b,t_a)$ might appear to be rather unpleasant, but in the following we shall show that it contains many interesting features. First we give an integral representation of one of the kernels appearing in (3.4),

$$K(q_{k+1},t_{k+1};q_k,t_k)$$

$$= e^{-\frac{i\tau}{\hbar}V(q_{k+1})}\, e^{\frac{i\hbar\tau}{2m}(\partial/\partial q_{k+1})^2}\,\delta(q_{k+1}-q_k)$$

$$=\frac{1}{h}\int_{-\infty}^{\infty} dp_k \exp\left[\frac{i\tau}{\hbar}\left\{p_k\left(\frac{q_{k+1}-q_k}{\tau}\right)-\frac{1}{2m}p_k{}^2-V(q_{k+1})\right\}\right]$$

Hence the virtual amplitude (3.4) written in the form of a functional integral, is

$$\Phi[q(t)] = \frac{1}{\hbar}\int dp_0 \int_F d[p(t)/h] \tag{3.10}$$

$$\cdot \exp\left\{\frac{i}{h}\int_{t_a}^{t_b} dt\left[p(t)\dot{q}(t)-\frac{1}{2m}p(t)^2-V[q(t)]\right]\right\},$$

where we have to note that

$$\int_F d[p(t)/h] = \lim_{n\to\infty}\int_{(n-1)}\!\!\cdots\int \prod_{j=1}^{n-1}\frac{1}{h}dp_k$$

just as in (3.5). The integrand in the exponent is just the Lagrangian $L=p\dot{q}-H$. Since $p\dot{q}-\frac{1}{2m}p^2=\frac{m}{2}\dot{q}^2-\frac{1}{2m}(p-m\dot{q})^2$, the virtual amplitude $\Phi[q(t)]$ is given by

$$\Phi[q(t)] = \exp\left[\frac{i}{\hbar}\int_{t_a}^{t_b} dt\, L[\dot{q}(t),q(t)]\right]$$

$$\cdot\frac{1}{h}\int dp_0 \int_F d[p(t)/h]\exp\left[-\frac{i}{2m\hbar}\int_{t_a}^{t_b} dt\,[p(t)-m\dot{q}(t)]^2\right]. \tag{3.11}$$

We see, therefore, that the quantity in the second line gives a functional integral representation of $N(t_b, t_a)$. It is easy to see that $p(t) = m\dot{q}(t)$ in the limit $h \to 0$.

Moreover, it is interesting to see that the factor

$$\int_F d[p(t)/h] \exp\left[\frac{i}{h} \int_{t_a}^{tb} dt\; p(t)\dot{q}(t) \right]$$

gives nothing but the delta functional $\mathscr{D}[q(t)]$ where the functional delta symbol $\mathscr{D}[q(t)]$ is such that

$$\int_F d[q(t)]\; \mathscr{D}[q(t)] = 1 \quad \text{and} \quad \mathscr{D}[q(t)] = 0 \text{ for } q(t) \neq 0 \tag{3.12}$$

and is defined as an infinite product of one-dimensional delta functions

$$\delta(qk) = \frac{\tau}{h} \int_{-\infty}^{\infty} df_k \quad e^{i\tau f_k q_k/\hbar}$$

That is

$$\mathscr{D}[q(t)] = \lim_{n\to\infty} \prod_{k=1}^{n-1} \delta(q_k)$$

$$= \lim_{n\to\infty} \underbrace{\int \cdots\cdots \int}_{(n-1)} \prod_{k=1}^{n-1} \frac{\tau}{h}\, df_k \exp\left[\frac{i}{\hbar} \sum_{k=0}^{n-1} \tau f_k q_k \right],$$

where we have assumed $q_0=0$. If we define a new variable p_k by

$$p_k = p_0 + \sum_{j=1}^{k} \tau f_j \;,$$

then we have $\prod_{k-1}^{n-1} dp_k = \prod_{k=1}^{n-1} \tau df_k$, and accordingly

$$\mathcal{D}[q(t)] = \int_F d[p(t)/h] \exp\left[\frac{i}{\hbar}\int_{t_a}^{t_b} dt\ f(t)q(t)\right] \quad (3.13)$$

In a continuous language the above variable transformation is rewritten as

$$p(t) = p(t_a) + \int_{t_a}^{t} ds\ f\ (s) \quad \text{and} \quad \dot{p}(t) = f(t)\ , (3.14)$$

and we might regard f(t) as the force acting on the particle.

Assuming $q(t_b) = 0$ we obtain after integration by parts

$$\mathcal{D}[q(t)] = \int_F d[p(t)/h] \exp\left[\frac{i}{\hbar}\int_{t_a}^{t_b} dt\ \dot{p}(t)q(t)\right]$$

$$= \int_F d[p(t)/h] \exp\left[-\frac{i}{\hbar}\int_{t_a}^{t_b} dt\ p(t)\dot{q}(t)\right]. (3.15)$$

Since the sign of the exponent can be changed in this case, we have proved the above statement.

If (3.14) is replaced by

$$p(t)=p(t_a)+\int_{t_a}^{t} ds\beta(s)f(s) \quad \text{and} \quad \dot{p}(t) = \beta(t)f(t)(3.16)$$

with a function $\beta(t)$, which will be defined later, then we have $\prod_{k=1}^{n-1} dpk/\beta(t_k) = \prod_{k=1}^{n-1}\tau df_k$ and accordingly

$$\mathcal{D}[q(t)] = \int_F d\,[p(t)/h\beta(t)]\ \exp\left[\frac{i}{\hbar}\int_{t_a}^{t_b} dt\ f(t)q(t)\right]$$

$$= \int_F d\,[p(t)/h\beta(t)]\ \exp\left[-\frac{i}{h}\int_{t_a}^{t_b} dt\ p(t)\,\frac{d}{dt}\,[q(t)/\beta(t)]\right]. \tag{3.17}$$

4. EXPANSION AROUND ACTUAL CLASSICAL PATH

The virtual amplitude $\Phi[q(t)]$ or the functional integrand (3.7) is expressed in terms of a fictitious path $q(t)$ connecting a pair of common end points (q_a, t_a) and (q_b, t_b). The actual classical path passing through these points, denoted by $q_c(t)$, is a solution to the Newtonian equation of motion

$$m\ddot{q}_c(t) + V^{(1)}[q(t)] = 0 \tag{4.1}$$

satisfying the terminal conditions $q_c(t_a) = q_a$ and $q_c(t_b) = q_b$. Here $V^{(k)}(q) = (d/dq)^k V(q)$. If we write

$$q(t) - q_c(t) = \xi(t) \tag{4.2}$$

and

$$W(q_c, \xi) = \sum_{k=3}^{\infty} \frac{1}{k!}\,\xi^k V^{(k)}(q_c), \tag{4.3}$$

then the potential function $V(q)$ can be expanded as

$$V(q) = V(q_c) + \xi V^{(1)}(q_c) + \tfrac{1}{2}\xi^2 V^{(2)}(q_c) + W(q_c, \xi).$$

The expansion of the Lagrangian $L(\dot{q}, q)$ thus reads

$$L(\dot{q},q) = \frac{m}{2}\dot{q}^2 - V(q)$$
$$= \frac{m}{2}\dot{q}_c^2 + m\dot{\xi}\dot{q}_c + \frac{m}{2}\dot{\xi}^2$$
$$-V(q_c) - \xi V^{(1)}(q_c) - \frac{1}{2}\xi^2 V^{(2)}(q_c) - W(q_c,\xi),$$

wherein $m\dot{\xi}\dot{q}_c - \xi V^{(1)}(q_c) = m\frac{d}{dt}(\xi\dot{q}_c)$ using (4.1). If we set

$$\xi(t) = \beta(t)\zeta(t) \qquad (4.4)$$

with $\dot{\xi} = \dot{\beta}\zeta + \beta\dot{\zeta}$, then

$$\frac{m}{2}\dot{\xi}^2 - \frac{1}{2}\xi^2 V^{(2)}(q_c)$$
$$= \frac{m}{2}\beta^2\dot{\zeta}^2 + \frac{m}{2}\left[2\zeta\dot{\zeta}\beta\dot{\beta} + \zeta^2\left\{\dot{\beta}^2 - \frac{1}{m}\beta^2 V^{(2)}(q_c)\right\}\right]$$
$$= \frac{m}{2}\beta^2\dot{\zeta}^2 + \frac{m}{2}\frac{d}{dt}(\zeta^2\beta\dot{\beta}) + \frac{m}{2}\zeta^2\left[\left\{\dot{\beta}^2 - \frac{1}{m}\beta^2 V^{(2)}(q_c)\right\} - \frac{d}{dt}(\beta\dot{\beta})\right]$$

We shall fix the function $\beta(t)$ in such a way that the quantity in the brackets vanishes. Thus we obtain a second-order linear differential equation for $\beta(t)$:

$$m\ddot{\beta}(t) + V^{(2)}(q_c)\beta(t) = 0. \qquad (4.5)$$

In view of (4.2) and (4.4) we have to assume the conditions

$$\zeta(t_a) = \zeta(t_b) = 0. \qquad (4.6)$$

The classical action appearing in the functional integrand, which is the Lagrangian function integrated with respect to time along a fictitious path $q(t)$, is simplified finally to

$$\int_{t_a}^{t_b} dt\, L[\dot{q}(t),q(t)] = S(q_b,t_b;q_a,t_a) \tag{4.7}$$
$$+\int_{t_a}^{t_b} dt\left[\frac{m}{2}\beta(t)^2\dot{\zeta}(t)^2 - W[q_c(t),\xi(t)]\right]$$

with the classical action function

$$S(q_b,t_b;q_a,t_a) = \int_{t_a}^{t_b} dt\, L[\dot{q}_c(t),q_c(t)]. \tag{4.8}$$

Here $\zeta(t)$ and accordingly $\xi(t)$ are still arbitrary except for the conditions (4.6).

Our second task in the present part is to establish the identity

$$\frac{\partial^2 S(q_b,t_b;q_a,t_a)}{\partial q_b \partial q_a} = -m\left[\beta(t_b)\beta(t_a)\int_{t_a}^{t_b} dt\,\beta(t)^{-2}\right]^{-1} \tag{4.9}$$

For this purpose we shall assume that the function $\zeta(t)$ has the form

$$\zeta(t) = \zeta_a + C\int_{t_a}^{t} ds\,\beta(s)^{-2} \tag{4.10}$$

with $\zeta(t_a) = \zeta_a$ and $\zeta(t_b) = \zeta_b$. We then have

$$\dot{\zeta}(t) = C\beta(t)^{-2}, \tag{4.11}$$

where the constant C is given by

$$\zeta_b - \zeta_a = C\int_{t_a}^{t_b} dt\,\beta(t)^{-2}.$$

Therefore, one obtains

$$\int_{t_a}^{t_b} dt\, \frac{m}{2}\beta(t)^2\dot{\zeta}(t)^2 = \frac{m}{2}\, c^2 \int_{t_a}^{t_b} dt\,\beta(t)^{-2}$$

$$= \frac{m(\zeta_b - \zeta_a)^2}{2\int_{t_a}^{t_b} dt\,\beta(t)^{-2}} \quad . \tag{4.12}$$

Here we have to note that $\zeta(t)$ and $\xi(t)$ for $t_b > t > t_a$ depend linearly on the terminal values ζ_a and ζ_b. Since the terminal values of the actual classical path $q_c(t)$ are already fixed as $q_c(t_a) = q_a$ and $q_c(t_b) = q_b$ and accordingly we have the terminal conditions

$$q(t_a) = q_a + \beta(t_a)\,\zeta_a \quad ,$$
$$q(t_b) = q_b + \beta(t_b)\,\zeta_b \quad ,$$

the fictitious path $q(t) = q_c(t) + \xi(t)$ for $t_b > t > t_a$ is determined uniquely by assuming the values of ζ_a and ζ_b or equivalently those of $q(t_a)$ and $q(t_b)$. The differential operator $\partial^2/\partial q(t_b)\partial q(t_a) = [\beta(t_b)\beta(t_a)]^{-1}\partial^2/\partial\zeta_b\partial\zeta_a$ operating on both sides of (4.7) eliminates the classical action function as well as the surface terms neglected therein. We thus have

$$\frac{\partial^2}{\partial q(t_b)\partial q(t_a)}\int_{t_a}^{t_b} dt\; L[\dot{q}(t), q(t)] + m\Big[\beta(t_b)\beta(t_a)\int_{t_a}^{t_b} dt\,\beta(t)^{-2}\Big]^{-1}$$

$$= -[\beta(t_b)\beta(t_a)]^{-1}\int_{t_a}^{t_b} dt\, \sum_{n=3}^{\infty} \frac{1}{n!}\, V^{(n)}[q_c(t)]\,\beta(t)^n \frac{\partial^2}{\partial\zeta_b\partial\zeta_a}\zeta(t)^n . \tag{4.13}$$

Now letting both ζ_a and ζ_b tend to zero, we have

$$\zeta(t) \to 0, \xi(t) \to 0 \text{ and } q(t) \to q_c(t). \qquad (4.14)$$

In this limit the righthand side of (4.13) vanishes and the lefthand side yields the required identity (4.9).

5. HOW TO PERFORM FUNCTIONAL INTEGRATION

In this final part of my lectures we shall investigate how to perform functional integration by combining the results obtained in earlier parts. In equation (3.7) the functional integrand was written in the form

$$\Phi[q(t)] = N(t_b,t_a) \exp\left[\frac{i}{\hbar}\int_{t_a}^{t_b} dt\ L[\dot{q}(t),q(t)]\right]$$

which represents a virtual amplitude for a fictitious path $q(t)$ connecting common end points (q_a,t_a) and (q_b,t_b). A functional integral representation of the normalization factor $N(t_b,t_a)$ was given in the second line of (3.11) as

$$N(t_b,t_a) = \frac{1}{h}\int dp_0 \int_F d[p(t)/h] \exp\left[-\frac{i}{2m\hbar}\int_{t_a}^{t_b} dt\ p(t)^2\right], \qquad (5.1)$$

where we have shifted the origin of each $p(t)$. the most convenient functional integral representation of the delta functional for the following developments will be

$$\mathcal{D}[\xi(t)] = \int_F d[p(t)/h\beta(t)] \exp\left[-\frac{i}{\hbar}\int_{t_a}^{t_b} dt\ p(t)\dot{\zeta}(t)\right] \qquad (5.2)$$

Here the $q(t)$ in (3.17) has been replaced by $\xi(t)=\beta(t)\zeta(t)$ in view of the considerations made in section 4, where we obtained the expansion (4.7)

$$\int_{t_a}^{t_b} dt\, L[\dot{q}(t),q(t)] = S(q_b,t_b;q_a,t_a) + \int_{t_a}^{t_b} dt\, \frac{m}{2}\beta(t)^2\dot{\zeta}(t)^2 - \int_{t_a}^{t_b} dt\, W[q_c(t),\xi(t)] .$$

If one compares all the four equations listed above it is tempting to conclude that a simple change of integration variables in (5.1): $p(t) \rightarrow p(t)/\beta(t)+m\beta(t)\dot{\zeta}(t)$ will yield (5.2) as well as to absorb the kinetic energy part in (4.7). This kind of simple transformation does not work however, as will be shown in the following.

For a new time parameter θ, which is related to the original time t by

$$\theta(t) = \int_0^t ds\, \alpha(s)^2 , \tag{5.3}$$

we have with the abbreviation $\alpha_k = \alpha(t_k)$

$$\Delta\theta_k = \theta(t_{k+1}) - \theta(t_k) = \int_{t_k}^{t_k+\tau} ds\, \alpha(s)^2$$

$$= \tau\alpha_k + \tau^2\alpha_k\dot{\alpha}_k + O(\tau^3) = [\alpha_k + \tau\dot{\alpha}_k + O(\tau^2)]\tau\alpha_k$$

$$= \tau\alpha_k\alpha_{k+1}[1 + O(\tau^2)] ,$$

so that

$$\left(\frac{m}{ih\tau}\right)^{n/2} = \prod_{k=0}^{n-1}\left[\frac{m\alpha_{k+1}\alpha_k}{ih\Delta\theta_k}\right]^{\frac{1}{2}}[1+O(\tau^2)]$$

$$= [1+O(n\tau^2)] \left[\frac{m\alpha_n\alpha_0}{ih\Delta\theta_0}\right]^{\frac{1}{2}} \prod_{k=1}^{n-1} \alpha_k \left[\frac{m}{ih\Delta\theta_k}\right]^{\frac{1}{2}}$$

$$= [1+O(\tau)] (\alpha_n\alpha_0)^{\frac{1}{2}} \frac{1}{h}\int dp_0 \exp\left[-\frac{i}{2m\hbar}\Delta\theta_0 p_0{}^2\right]$$

$$\cdot \prod_{k=1}^{n-1} \frac{1}{h} \alpha_k dp_k \exp\left[-\frac{i}{2m\hbar}\Delta\theta_k p_k{}^2\right].$$

If we set $\alpha(t) = \beta(t)^{-1}$ and take the limit $n \to \infty$, we obtain

$$N(t_b,t_a) = [\beta(t_b)\beta(t_a)]^{-\frac{1}{2}} \frac{1}{h}\int dp_0 \int d[p(t)/h\beta(t)] \cdot$$

$$\exp\left[-\frac{i}{2m\hbar}\int_{t_a}^{t_b} dt\,\beta(t)^{-2}p(t)^2\right]. \tag{5.4}$$

The problem was the first factor on the righthand side, which we have already seen in (4.9).

By substituting $p(t)+m\beta(t)^2\dot{\zeta}(t)$ for $p(t)$ in the above we see that the functional integrand $\Phi[q(t)]$ simplifies

$$\Phi[q(t)] = \exp\left[\frac{i}{\hbar}\left[S(q_b,t_b;q_a,t_a)-\int_{t_a}^{t_b} dt\, W(q_c,\xi)\right]\right]$$

$$\cdot [\beta(t_b)\beta(t_a)]^{-\frac{1}{2}} \frac{1}{h}\int dp_0 \int_F d[p(t)/h\beta(t)] \tag{5.5}$$

$$\cdot \exp\left[-\frac{i}{\hbar}\int_{t_a}^{t_b} dt\left[\frac{1}{2m}\beta(t)^{-2}p(t)^2+p(t)\dot{\zeta}(t)\right]\right],$$

where the factors corresponding to the delta functional (5.2), appear explicitly. In order first to perform the

p_o-integration we recall the transformation (3.16), that is,

$$p(t) = p_o + I(t) \quad \text{with} \quad I(t) = \int_{t_o}^{t} ds\beta(s)f(s) \qquad (5.6)$$

and $\dot{p}(t) = \beta(t)f(t)$. If we write

$$B = \int_{t_a}^{t_b} d\theta(t), \quad J_1 = \int_{t_a}^{t_b} I(t)d\theta(t), \quad J_2 = \int_{t_a}^{t_b} I(t)^2 d\theta(t), \qquad (5.7)$$

the second and third lines in (5.5) are transformed after the p_o-integration into

$$\left[\frac{m}{ih\beta(t_b)\beta(t_a)B}\right]^{\frac{1}{2}} \int_F d[p(t)/h\beta(t)]$$

$$\cdot \exp \frac{i}{\hbar} \left[\frac{1}{2m} \left(\frac{1}{B} J_1{}^2 - J_2\right) + \int_{t_a}^{t_b} dt\ f(t)\xi(t)\right], \qquad (5.8)$$

where $m/\beta(t_b)\beta(t_a)B = -\partial^2 S/\partial q_b \partial q_a$ according to (4.9) Here we must note that both J_1 and J_2 are functionals in f(t), (the former being linear and the latter quadratic), and moreover that f(t) multiplying $\exp\left[i/h \int dt\ f(t)\xi(t)\right]$ is equivalent to a functional differential operator $-i\hbar\delta/\delta\xi(t)$. Hence on account of (3.17) the above (5.8) can further be rewritten as

$$\left[\frac{i}{h} \frac{\partial^2 S}{\partial q_b \partial q_a}\right]^{\frac{1}{2}} \exp\left[i\hbar\ D[\delta/\delta\xi(t)]\right]\mathcal{D}[\xi(t)] \qquad (5.9)$$

with

$$2mD[\delta/\delta\xi(t)] = \frac{1}{B} J_{1,op}{}^2 - J_{2,op}\ .$$

In view of (2.8) the functional integrand $\Phi[q(t)]$ thus assumes its final form:

$$\Phi[q(t)] = K_c(q_b,t_b;q_a,t_a)\ \exp\left[-\frac{i}{\hbar}\int_{t_a}^{t_b} dt\ W[q_c(t),\xi(t)]\right]$$

$$\cdot\ \exp\left\{i\hbar D[\delta/\delta\xi(t)]\right\}\mathcal{D}[\xi(t)] \qquad (5.10)$$

where $\xi(t)$ denotes the deviation of a fictitious path $q(t)$ from the actual classical path $q_c(t)$: $\xi(t)=q(t)-q_c(t)$.

Now extending the property of the one-dimensional delta function we easily see that

$$F[\xi]\frac{\delta}{\delta\xi}\mathcal{D}[\xi] = -\mathcal{D}[\xi]\left[\frac{\delta}{\delta\xi}F[\xi]\right]_{\xi=0}$$

for a reasonable functional $F[\xi]$. Hence (5.10) is equivalent to

$$\Phi[q] = K_c(q_b,t_b;q_a,t_a)\mathcal{D}[\xi]Z[q_c] \qquad (5.11)$$

with

$$Z[q_c] = \left[\exp\left\{i\hbar D[\delta/\delta\xi]\right\}\cdot\exp\left\{\frac{1}{i\hbar}\int_{t_a}^{t_b} dt\ W[q_c,\xi]\right\}\right]_{\xi=0}, \qquad (5.12)$$

where we have to note that $D[\delta/\delta\xi]$ is quadratic in $\delta/\delta\xi$. Equation (5.11) is integrated functionally to give the final result

$$K(q_b,t_b;q_a,t_a) = \int d[q]\ \Phi[q] = K_c(q_b,t_b;q_a,t_a)Z[q_c] \qquad (5.13)$$

Now we have to investigate the nature of the correction factor (5.12). The quantity in the brackets is expanded in powers of h as

$$\sum_{j,k=0}^{\infty} \frac{1}{j!k!} (i\hbar)^{j-k} \left\{D[\delta/\delta\xi]\right\}^{j} \left\{\int_{t_a}^{t_b} dt\ W(q_c, \xi)\right\}^{k}$$

Since the $W(q_c,\xi)$ is at least cubic in ξ, nonvanishing contributions for $\xi=0$ come only from the terms with $2j\geq 3k$ or $n=j-k\geq k/2\geq 0$. Therefore,

$$Z[q_c] = 1 + \sum_{n=1}^{\infty} (i\hbar)^n Z_n[q_c] \tag{5.14}$$

with

$$Z_n[q_c] = \sum_{k=1}^{2n} \frac{1}{k!(k+n)!} \left[D[\delta/\delta\xi]^{k+n} \left\{\int_{t_a}^{t_b} dt\ W(q_c,\xi)\right\}^{k} \right]_{\xi=0}. \tag{5.15}$$

From (5.10) we see that, for an identically vanishing $W(q_c,\xi)$ corresponding to $d^3V(q)/dq^3=0$, the classical limit of the virtual amplitude is given by

$$\lim_{h\to 0} \Phi[q(t)] = K_c(q_b,t_b;q_a,t_a)\mathcal{D}[q(t)-q_c(t)] \ . \tag{5.16}$$

But even though $W(q_c,\xi)$ does not vanish, the above arguments confirm that the net effect of the formal limiting procedure $h\to 0$ is still given by (5.16)

The combined equations (5.13) and (5.14) afford a kind of W.K.B. approximation to the kernel $K(q_b,t_b;q_a,t_a)$. But we have to regret that all the above derivations presuppose the knowledge of classical functions $q_c(t)$ and $\beta(t)$. This point makes it hard to see the content of our exact results obtained by performing the functional integration.

The generalization of the above procedures to the case of many degrees of freedom and to field theory are contained in ref. [5].

REFERENCES

1) R.P. Feynman and A.R. Hibbs, "Quantum Mechanics and Path Integrals", (McGraw-Hill Book Co., New York, 1965).

2) M. Kac, "Probability and Related Topics in Physical Sciences", (Interscience Publishers, Inc., New York, 1959)

3) R.P. Feynman, Rev.Mod.Phys. 20, 367 (1948).

4) C. Morette, Phys. Rev. 81, 848 (1951); I.M. Gel'fand, and A. M. Yaglom, J.Math.Phys. 1, 48 (1960); S.G. Brush, Rev.Mod.Phys. 33, 79 (1961); S.S. Schweber, J.Math.Phys. 3, 831 (1962); C. Garrod, Rev.Mod. Phys. 38, 483 (1966); G. Rosen, Phys.Fluids 10, 2614 (1967); L. Schulman, Phys.Rev. 176, 1558 (1968); W.E. Brittin and W.R. Chappell, J.Math.Phys. 10, 661, (1969); C. Blomberg, J.Math.Phys. 10, 675 (1969).

5) I. Fujiwara, Det Kgl.Norske Vid.Selsk.Forh. 36, 57, 60, 72, 77, 81, 86, 97, 102, 107, 111, 115, 121, (1963); 39, 73, 80, (1966); 40, 37 (1967).

6) I. Fujiwara and P.Chr. Hemmer, Det Kgl.Norske Vid.Selsk. Forh. 36, 46, 49, 53 (1963).

7) P.A.M. Dirac, Phys.Z.d.Sovj.U. 3, 64 (1933).

8) I. Fujiwara, Prog.Theor.Phys. 21, 902 (1959).

9) I. Fujiwara, Prog.Theor.Phys. 38, 1 (1967).

FIELD EQUATIONS AND FORM-INVARIANT RENORMALIZATION

E.R. CAIANIELLO

Laboratorio di Cibernetica del C.N.R.
Arco Felice
Napoli

1. OUTLINE OF LECTURES

The aim of these lectures is to give a concise presentation of the techniques developed by the author[1)] and his collaborators[2-5)], notably M. Marinaro and F. Guerra, in order to obtain a systematic treatment of the ultra violet infinities which appear as soon as one attempts to solve field theoretic equations.

The problem itself is too well known to bear repetition. We wish, here, only to emphasise some of the points at which our method differs from the traditional ones. All the work carried out on renormalization by other authors has been essentially with perturbation expansions. Even when field equations are written down, starting from Lagrangians with counter terms (and therefore apparently divergence free) no guarantee is offered that divergences do not appear when methods other than the perturbative expansion are used for their solution.

Our technique based on the system of finite part integrals (F.P.I.) is, instead, independent of the method of solution and of the approximations used. It gives a precise meaning to otherwise meaningless equations connecting distributions by appropriately defining the F.P.I. (which must replace the ordinary integral), and, if the prescription for the evaluation of the F.P.I. is changed (as can be done in an infinity of ways), this is explicitly shown to amount to the standard renormalization of masses and charges. There are no infinite quantities, only arbitrary ones, as are, in fact, the "bare" masses and charges of the conventional method. This behaviour occurs only if the theory itself satisfies appropriate conditions which are identical to those of Dyson for renormalizability.

At all stages a precise book-keeping of all quantities subtracted and recombined is made possible by an extensive use of combinatorics, which is also essential for attacking, in a consistent manner, the rather delicate points which arise when handling such problems (e.g. the "point-loop" ambiguity). The first lecture will thus be devoted to a review of the combinatorial results which are more relevant for our purposes, the second to writing down equations for Green functions (or better, as we shall see, "propagators") of the theory. (Electrodynamics and neutral scalar mesons with $g\varphi^4$ coupling will be recurring examples.) Then we shall point in two different directions: the first, ignoring all analytical troubles in integrals and using coarse numerical examples, will be to exhibit the kind of behaviour one may expect from such equations, namely solutions that appear to be non-analytic in the coupling constant (actually of Hadamard class 2); the second will be to give, as promised, a well-defined meaning to these equations, introducing F.P.I's and showing how these work. It will also yield a satisfactory definition of the renormalization group. To avoid too heavy manipulations this will be shown only for simple examples, the general cases being treated in the references.

The sum total of our work, from a practical point of view, will be that, once one is sure that a theory is renormalizable, by using an appropriate F.P.I., renormalization can be completely forgotten about, no matter whether one solves the field equations perturbatively or not.

2. COMBINATORICS

We consider a field theory in which the equations of motion are known and we look for the algebraic tools to handle these equations. The particular theory to which we shall refer in this section is electrodynamics.

There are two algorithms which we must consider: the pfaffians and the hafnians. The pfaffians, well known for a long time, appear in a natural way when we consider fermion fields. Correspondingly, when we consider a boson field the appropriate algebraic tool is the hafnian which has been introduced for the first time by the author in connection with this kind of problem.

Let $x_1, x_2, \ldots, x_n$ be the elements of an algebra. For them we define two different kinds of product

(Grassmann algebra)

$$x_h \wedge x_k = - x_k \wedge x_h \quad (x_h \wedge x_h = 0) \tag{2.1}$$

(Clifford Algebra)

$$x_h \mathbin{\dot\wedge} x_k = - x_k \mathbin{\dot\wedge} x_h + 2\delta_{hk} \quad (x_h \mathbin{\dot\wedge} x_h = 1) \tag{2.2}$$

2.1 Determinants

Let

$$\omega_h = \sum_{i=1}^{h} \alpha_{hi}\, x_i \tag{2.3}$$

be a linear form in the elements x_i $(i = 1,2,\ldots, n)$.

We are interested in finding a convenient expression of products like:

$$\omega_1 \wedge \omega_2 \, .$$

We have by virtue of (2.1):

$$\omega_1 \wedge \omega_2 = \sum_{i \neq j} \alpha_{1i}\, \alpha_{2j}\, x_i \wedge x_j = \sum_{i<j} \begin{vmatrix} \alpha_{1i} & \alpha_{1j} \\ \alpha_{2i} & \alpha_{2j} \end{vmatrix} x_i \wedge x_j \, .$$

We see that determinants have been introduced in quite a natural way.

In the general case of products like:

$$\Pi = \omega_1 \wedge \omega_2 \ldots \wedge \omega_k \, ,$$

we have, by an easy extension, using the Sylvester's notation for determinants, i.e.:

$$\begin{pmatrix} 1 & 2 & \dots & k \\ i_1 & i_2 & \dots & i_k \end{pmatrix} = \begin{vmatrix} \alpha_{1i_1} & \alpha_{1i_2} & \dots & \alpha_{1i_k} \\ \alpha_{2i_1} & \alpha_{2i_2} & \dots & \alpha_{2i_k} \\ \dots & \dots & \dots & \dots \\ \alpha_{ki_1} & \alpha_{ki_2} & \dots & \alpha_{ki_k} \end{vmatrix} \tag{2.4}$$

$$\Pi = \sum_{i_1 < i_2 .. i_k} \begin{pmatrix} 1, & 2 & \dots & k \\ i_1 & i_2 & \dots & i_k \end{pmatrix} x_{i_1} \wedge x_{i_2} \wedge \dots x_{i_k}$$

$$= \frac{1}{k!} \sum_{i_1, i_2 .. i_k = 1}^{n} \begin{pmatrix} 1, & 2 & \dots & k \\ i_1 & i_2 & \dots & i_k \end{pmatrix} x_{i_1} \wedge x_{i_2} .. x_{i_k}.$$

If $k = n$ there is only one permutation $i_1 = 1$, $i_2 = 2, \dots, i_n = n$ possible and we get:

$$\omega_1 \wedge \omega_2 \dots \wedge \omega_n = \begin{pmatrix} 1 & 2 & \dots & n \\ 1 & 2 & \dots & n \end{pmatrix} x_1 \wedge x_2 \wedge \dots x_n \tag{2.5}$$

Equations (2.3) to (2.5) could be used in order to define determinants, and all theorems about determinants could be deduced immediately by them.

For instance, one could prove the following expansion of determinants by minors of the rows 1 and 2:

$$\begin{pmatrix} 1 & 2 & \dots & n \\ 1 & 2 & \dots & n \end{pmatrix} = \sum_{\substack{i_1 < i_2 \\ i_3 < .. i_n}} (-1)^p \begin{pmatrix} 1 & 2 \\ i_1 & i_2 \end{pmatrix} \begin{pmatrix} 3 & \dots & n \\ i_3 & \dots & i_n \end{pmatrix}$$

where p is the parity of $i_1, i_2, \dots i_n$ with respect to their natural order $1, 2, \dots, n$.

We have

$$\omega_1\wedge\omega_2\ldots\wedge\omega_n = (\omega_1\wedge\omega_2)\wedge(\omega_3\wedge\ldots\omega_n)$$

$$=\left[\sum_{i_1<i_2}\begin{pmatrix}1 & 2\\ i_1 & i_2\end{pmatrix} x_{i_1}\wedge x_{i_2}\right]\wedge\left[\sum_{i_2<\ldots i_n}\begin{pmatrix}3 & .. & n\\ i_3 & .. & i_n\end{pmatrix} x_{i_2}\wedge \ldots x_{i_n}\right]$$

$$= \sum_{\substack{i_1<i_n\\ i_3<..i_n}} (-1)^p \begin{pmatrix}1 & 2\\ i_1 & i_2\end{pmatrix}\begin{pmatrix}3 & \ldots & n\\ i_3 & \ldots & i_n\end{pmatrix} x_1\wedge x_2 \ldots\wedge x_n \ .$$

By comparing the last expression with (2.5) we get the result.

2.2 Pfaffians

The pfaffians can be introduced in several ways: the most useful are the following:

Start from a quadratic external form:

$$\Omega = \frac{1}{2!}\sum_{h,k=1}^{n} \alpha_{hk} x_h\wedge x_k = \sum_{h<k}\alpha_{hk}\, x_h\wedge x_k \ ,$$

with

$$\alpha_{hk} = -\alpha_{kh} \ .$$

Its (Grassmann) powers are:

$$\Omega_1\wedge\Omega_2\ldots\wedge\Omega_i = (\Omega)^i\wedge .$$

We consider $(\Omega)^2\wedge$:

$$\Omega\wedge\Omega = \sum_{h_1<h_2}\ \sum_{h_3<h_4}\alpha_{h_1h_2}\alpha_{h_3h_4} x_{h_1}\wedge x_{h_2}\wedge x_{h_3}\wedge x_{h_4} =$$

$$= 2!\sum_{h_1<h_2<h_3<h_4}(\alpha_{h_1h_2}\alpha_{h_3h_4} - \alpha_{h_1h_3}\alpha_{h_2h_4} + \alpha_{h_1h_4}\alpha_{h_2h_4}) \cdot$$

$$\cdot\ x_{h_1}\wedge x_{h_2}\wedge x_{h_3}\wedge x_{h_4} \ .$$

From now on, when $h < k$ we shall use the notation $\alpha_{hk} = (hk)$. Therefore:

$$\Omega \wedge \Omega = 2! \sum_{h_i < h_2 < h_3 < h_4} \Big[(h_1 h_2)(h_3 h_4) - (h_1 h_3)(h_2 h_4) + (h_1 h_4)(h_2 h_3) \Big] \, x_{h_1} \wedge \cdots \wedge x_{h_4} \, ,$$

$$= 2! \sum_{h_1 < h_2 < h_3 < h_4} (h_1 h_2 h_3 h_4) x_{h_1} \wedge x_{h_2} \wedge x_{h_3} \; x_{h_4} \, ,$$

where we have introduced the symbol:

(which we shall call pfaffian):

$$(1\ 2\ 3\ 4) = (1\ 2)(3\ 4) - (1\ 3)(2\ 4) + (1\ 4)(2\ 3) \, .$$

Therefore in the general case the definition of a pfaffian will be given by:

$$(\Omega)^{i}_{\wedge} = \ell! \sum_{h_1 > h_2 \ldots < h_{2\ell}} (h_1 h_2 \ldots h_{2\ell}) x_{h_1} \wedge x_{h_2} \ldots x_{h_2} \, . \tag{2.6}$$

For $n = 2m$ the maximum value of ℓ is m, because the Grassmann product of two equal elements is zero.

For the same reason when $n = 2m + 1$ the maximum value of ℓ is m.

In order to find the expansion rule for pfaffians we apply the definition (2.6) and we get in complete analogy with the case of determinants:

$$(1\ 2\ \ldots\ 2m) = \sum_{h=2}^{2m} (-1)^{h} (1\ h)(2\ \ldots\ h-1),\ h+1\ \ldots\ 2m). \tag{2.7}$$

At this point it is convenient to represent a pfaffian as a triangular array, that is:

$$(1\ 2\ \ldots\ 2m) = \left| \begin{matrix} (1\ 2)(1\ 3)\ \ldots\ (1\ \ 2n) \\ (2.3)\ \ldots\ (2\ \ 2n) \\ (2n-1,\ 2n) \end{matrix} \right|$$

In fact a pfaffian (in analogy with a determinant) can be expanded by the elements of one of its lines: "line h" is that one, in the triangular array, which contains all the elements carrying the index h, regardless of whether h is the first or second index. The element (hk) belongs to the lines h and k; crossing these lines out, what is left is a new pfaffian, the "minor" of (hk). The rule for the expansion is then: "A pfaffian is equal to the sum of the product of each element (hk) of a prefixed line by its minor, each term in the sum being given the sign $(-1)^{h+k+1}$." For instance if we fix the line 1, this rule leads to (2.7).

By repeated application of equation (2.7) we obtain:

$$(1\ 2\ \ldots\ 2m) = \sum{}'(-1)^p (i_1\ i_2)(i_3\ i_4)\ \ldots\ (i_{2n-1} i_{2n}).$$

$\sum'$ means here, and <u>throughout the following</u>, summation over all the permutations $i_1\ i_2\ \ldots\ i_{2n}$ of $1,2,\ldots\ 2n$ which satisfy the limitations:

$$i_1 < i_2 \quad i_3 < i_4\ \ldots\ i_{2n-1} < i_{2n} \quad i_1 < i_3 < i_5\ \ldots < i_{2n-1}\ .$$

A fundamental theorem on pfaffians (that we will not prove) is the following

$$\begin{vmatrix} 0 & (1\ 2) & (1\ 3) & \ldots & (1,\ 2m) \\ -(1\ 2) & 0 & (2\ 3) & & \cdot \\ -(1\ 3) & -(2\ 3) & 0 & & \cdot \\ \cdot & \cdot & \cdot & \cdot & \cdot \\ -(1,2m) & \cdot & \cdot & \cdot & 0 \end{vmatrix} = (1\ 2\ \ldots\ 2m)^2.$$

Exercise. Consider a product of 4 x 4 matrices:

$$P_1 P_1 \dots P_{2m} ,$$

where $P_r = \gamma^\mu p_\mu^{(r)}$ and $\gamma^\mu \gamma^\nu + \gamma^\nu \gamma^\mu = 2\delta^{\mu\nu}$.

Show that

$$\frac{1}{4} Sp(P_1 P_2 \dots P_{2m}) = (1\ 2 \dots 2m),$$

where (hk) is the scalar product $\vec{p}^{(h)} \cdot \vec{p}^{(k)}$.

If we count how many terms there are in a pfaffian we find: number of terms of

$$(1\ 2\ \dots\ 2m) = (2m-1)!! = \frac{(2m)!}{2^m m!} .$$

Now we state (without proof) the main theorem about pfaffians. This is an extension of what we know about determinants. This theorem which contains as a particular case (2.7) is expressed by the following identity:

$$(1\ 2 \dots 2m) = \sum_{r=0}^{q/2} (-1)^{\left(\frac{q-2r}{2}\right)} \sum_{c_r} (-1)^p (k_1' k_2' \dots k_{2r}') \cdot \begin{pmatrix} k_1'' & k_2'' & \dots & k_s'' \\ \ell_1'' & \ell_2'' & \dots & \ell_s'' \end{pmatrix} (\ell_1' \ell_2' \dots \ell_{2t}') , \tag{2.7'}$$

where the expansion is made with respect to $q = 2r + s$ indices $k_1 k_2 \dots k_q$ arbitrarily chosen in the pfaffian.

(Of course $s + 2t + q = 2m$)

$\sum_{c_r}$ is extended to all the $\binom{q}{2r}\binom{2m-q}{2t}$ permutations:

$$k_1' < k_2' \dots < k_{2r}' \ ; \quad k_1'' < k_2'' \dots < k_s''$$

of the indices $k_1 k_2 \dots k_q$, and

$$\ell_1'' < \ell_2'' \dots < \ell_s'' \; ; \quad \ell_1' < \ell_2' \dots < \ell_{2t}'$$

of the indices $\ell_1 \ell_2 \dots \ell_{2m-q}$.

2.3 Relation between Grassmann's Algebra and Clifford's Algebra

The relation between the two algebras is given by the following formula which we do not prove:

$$\omega_1 \wedge \omega_2 \wedge \dots \wedge \omega_h = \sum_{r=0}^{[k/2]} \sum_{c_r} (-1)^p \, (h_1 h_2 \dots h_{2r}) . \\ \cdot \omega_{\ell_1} \wedge \omega_{\ell_2} \dots \omega_{\ell_{k-2r}} ,$$

where $(h_1 h_2) = \sum_i \alpha_{h_1 i} \alpha_{h_1 i}$, etc.

$$h_1 < h_2 \dots < h_{2r} \; ; \quad \ell_1 < \ell_2 < \dots < \ell_{k-2r}$$

is a permutation of 1 2 ... k with parity p; $\sum_{c_r}$ is extended over all such permutations with fixed r. $[k/2]$ is the highest integer contained in $k/2$. In the case $k = 2$ (2.8) gives:

$$\omega_1 \wedge \omega_2 = (1\ 2) + \omega_1 \wedge \omega_2 .$$

This formula is essentially analogous to Wick's theorem on field operators:

$$T(AB) = :AB: + N(AB) ,$$

where :AB: represents the contracted factors.

The Clifford product that we have defined is just the kind of product which appears when we consider creation and annihilation operators of the Fermion field. Therefore determinants and pfaffians are the appropriate mathematical tools for the Fermion field.

We can define forms in the elements $x_1\ x_2\ \dots\ x_h$ with the rules:

$$x_h \sqcap x_k = \begin{cases} x_k \sqcap x_h & h \neq k \\ 0 & h = k \end{cases}$$

and

$$x_h \boxdot x_k = x_k \boxdot x_h \qquad h \neq k$$
$$x_h \boxdot x_h = 1$$

which corresponds to Grassmann product and Clifford product respectively.

Entirely similar results can be obtained from these commutation rules. And the only difference will be, in all the cases, the change of the sign − into + . In this way we will obtain the appropriate mathematical tools for the boson field.

2.4 Permanents

The counterpart of a determinant is a permanent. A "permanent" is quite similar to a determinant, the only difference being that each term of its development is taken with + sign. It is well known that the antisymmetrized product of N_o wave functions $u^{(1)}, u^{(2)} \dots u^{(N_o)}$ is the determinant:

$$\begin{pmatrix} u^{(1)} & u^{(2)} & \dots & u^{(N_o)} \\ 1 & 2 & & N_o \end{pmatrix} = \begin{vmatrix} u^{(1)}(x_1) & u^{(1)}(x_2) & \dots & u^{(1)}(x_{N_o}) \\ u^{(2)}(x_1) & u^{(2)}(x_2) & \dots & u^{(2)}(x_{N_o}) \\ \cdot & \cdot & \cdot & \cdot \\ u^{(N_o)}(x_1) & u^{(N_o)}(x_2) & \dots & u^{(N_o)}(x_{N_o}) \end{vmatrix} \qquad (2.9)$$

with $(u^{(h)}{}_k) = u^{(h)}(x_k)$ and the symmetrized product of P_0 wave functions is the permanent:

$$\begin{bmatrix} z^{(1)} & z^{(2)} & \dots & z^{(p_0)} \\ 1 & 2 & & p_0 \end{bmatrix} = \begin{vmatrix} z^{(1)}(x_1) & z^{(1)}(x_2) & \dots & z^{(1)}(x_{p_0}) \\ z^{(2)}(x_1) & z^{(2)}(x_2) & \dots & z^{(2)}(x_{p_0}) \\ z^{(p_0)}(x_1) & z^{(p_0)}(x_2) & \dots & z^{(p_0)}(x_{p_0}) \end{vmatrix}_+ \qquad (2.10)$$

with $[z^{(h)}{}_k] = z^{(h)}(x_k)$.

2.5 Hafnians

In complete analogy, the counterpart of a pfaffian is a new algorithm that we will call a hafnian.

We define a hafnian with the formula

$$[1\ 2\ \dots\ 2n] = \sum{}' [i_1 i_2][i_3 i_4]\dots[i_{2n-1} i_{2n}]. \qquad (2.11)$$

Square brackets are used, consistently, for permanents and hafnians, round brackets for determinants and pfaffians.

For instance, for n = 2, (2.11) gives:

$$[1\ 2\ 3\ 4] = [1\ 2][3\ 4] + [1\ 3][2\ 4] + [1\ 4][2\ 3].$$

The triangular array is useful also for the hafnians:

$$[1\ 2\ \dots\ 2n] = \begin{vmatrix} [1\ 2] & [1\ 3] & \dots & [1\ ,\ 2n] \\ & [2\ 3] & & [2\ ,\ 2n] \\ & & & \vdots \\ & & & [2n-1,\ 2n] \end{vmatrix}_+$$

2.6 Application to Electrodynamics

We wish to show how this formalism leads to a compact expression for the n-th order term of the perturbative expansion of Dyson's S-matrix.

In electrodynamics, with the usual meaning of the symbols, we have the expansion of Dyson's U matrix (in the interaction representation):

$$U(t_f, t_i) = \sum_{n=0}^{\infty} \frac{\lambda^n}{n!} \int_{t_i}^{t_f} d^4x_1 \int_{t_i}^{t_f} d^4x_2 \ldots \int_{t_i}^{t_f} d^4x_N \,.$$

$$\cdot\, T\left(\prod_{i=1}^{N} \psi_{\alpha_i}(x_i)\psi_{\beta_i}(x_i)A_{\mu_i}(x_i)\gamma^{\mu_i}_{\alpha_i\,\beta_i}\right).$$

We want to find the transition amplitude between final and initial states $|f\rangle$ and $|i\rangle$.

Let us calculate first the vacuum-vacuum transition amplitude:

$$|f\rangle = |i\rangle = |0\rangle, \quad M_{oo} = k_{oo} = \langle 0|U|0\rangle.$$

In the interaction representation the fields A_μ and ψ commute. Therefore we have to evaluate separately:

$$F = \langle 0|T(\psi_{\alpha_1}(x_1)\psi_{\beta_1}(x_1)\psi_{\alpha_2}(x_2)\psi_{\beta_2}(x_2)\ldots$$

$$\ldots\psi_{\alpha_N}(x_N)\psi_{\beta_N}(x_N))|0\rangle.$$

$$B = \langle 0|T(A_{\mu_1}(x_1)A_{\mu_2}(x_2)\ldots A_{\mu_N}(x_N))|0\rangle.$$

We start with the calculation of B. Call, for short, Π the product $A_{\mu_1}(x_1)A_{\mu_2}(x_2)\ldots A_{\mu_N}(x_N)$.

Wick's theorem states that:

$$T(\Pi) = :\Pi: + \sum :\Pi^{(2)}: + \sum :\Pi^{(4)}: + \ldots , \qquad (2.12)$$

where $\Pi^{(2h)}$ denotes any of the values Π assumes when h pairs of operators are contracted, and $\sum \Pi^{(2h)}$ denotes summation over all such values, for given h. It is then immediately seen that the only term of the expansion (2.12) which contributes to B is the term with all the A_μ contracted, i.e., with $h = n/2$. Therefore we see that only terms with even n will contribute to the perturbative expansion.

Then

$$B = \sum :\Pi^{(n)}: . \qquad (2.13)$$

By comparing (2.11) with (2.13) we get:

$$B = [1\ 2\ \ldots\ 2n] ,$$

where $[hk] = \langle 0 | T(A_{\mu_h}(x_h) A_{\mu_k}(x_k)) | 0 \rangle = \frac{1}{2} \delta_{\mu_h \mu_k} D_F(x_h - x_k)$.

Denote now the product of the operators of the fermion field by

$$p = 1\ 2\ 3\ \ldots\ 2n = \bar{\psi}_{\alpha_1}(x_1)\psi_{\beta_1}(x_1) \ldots \bar{\psi}_{\alpha_n}(x_n)\psi_{\beta_n}(x_n),$$

where $h = 2(h) - 1$ stands for the operator $\bar{\psi}_{\alpha_{(h)}}(x_{(h)})$, $h = 2(h)$ for $\psi_{\beta_{(h)}}(x_{(h)})$ $((h) = 1, 2, \ldots, n)$.

We denote with (α) the maximum integer contained in $(\alpha + 1)/2$.

In the same way as for the case of the boson field, we get: (remembering that $:\psi\psi: = :\bar{\psi}\bar{\psi}: = 0$)

$$F = (1\ 2\ \ldots\ 2n) , \qquad (2.14)$$

where

$$(hk) = 0 \quad \text{for } h, k \text{ both even or both odd}$$
$$(hk) = \tfrac{1}{2} S^F_{\beta(k)\alpha(k)}(x_{(h)} - x_{(h)}) \quad \text{for H odd, k even}$$
$$(hk) = -\tfrac{1}{2} S^F_{\beta(h)\alpha(k)}(x_{(h)} - x_{(k)}) \quad \text{for h even, k odd.}$$

If we use the fundamental expansion theorem (2.7′) we can see that because of (hk) = 0 for h,k both even or both odd, the pfaffian reduces to a determinant.

Collecting all the results, one finds:

$$k_o = M_{oo} = \sum_{n=0}^{\infty} \frac{\lambda^{2n}}{(2n!)} \int d1 \sum_{\alpha_1 \beta_1 \mu_1} \gamma^{\mu_1}_{\alpha_1 \beta_1} \cdots$$

$$\cdots \int d(2n) \sum_{\alpha_{2n} \beta_{2n} \mu_{2n}} \gamma^{\mu_{2n}}_{\alpha_{2n} \beta_{2n}} \begin{pmatrix} 1 & 2 & \ldots & 2n \\ 1 & 2 & \ldots & 2n \end{pmatrix} [1 \; 2 \; \ldots \; 2n]. \tag{2.15}$$

This formula gives automatically the contribution of all Feynman diagrams without external lines.

For instance, in the fourth order, expanding the determinant and the hafnian one obtains, among other terms,

$$\int d1 \sum_1 \gamma^1 \int d2 \sum_2 \gamma^2 \cdots$$

$$\cdots \int d4 \sum_4 \gamma^4 \; (1\;2)(2\;3)(3\;4)(4\;1)[1\;2][3\;4]$$

which clearly gives the contribution of the graph:

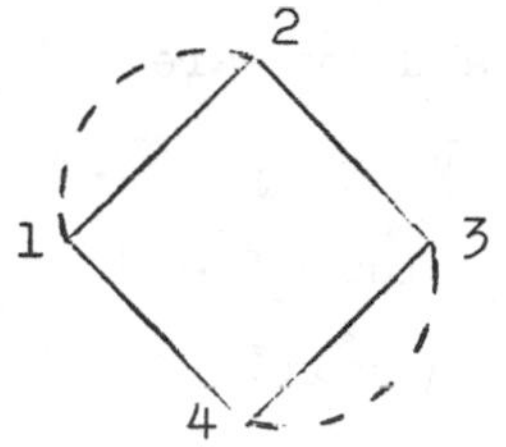

A remarkable feature of expansion (2.15) is that the contribution of fermions is separated from the contribution of bosons.

2.7 The General Perturbative Expansion

So far we have obtained the perturbative expansion of k_o, i.e., $\langle 0|U|0\rangle$. In essentially the same way, it is possible to obtain the element M_{FI} of the U matrix which describes the process in which

n electrons, n positrons, a photons

are destroyed and

p electrons, q positrons, b photons

are created, in states specified by some prescribed wave functions. Clearly n+q = m + p; we call N_o this number, and set $P_o = a + b$.

We will give, without proof, a general formula which expresses M_{FI} in terms of the wave functions of the initial and final states (with the due properties of symmetry and normalization) and of a "kernel", whose form depends only upon N_o and P_o.

Let $u_\alpha(x)$ $(x = x_1, x_2, x_0)$ be the wave function of an electron, $v_\beta(x)$ that of a positron, $\zeta_\mu(t)$ $(t \equiv t_1, t_2, t_3, t_0)$ that of a photon. (These functions being solutions of the free fields equations).

The adjoints of u and v are:

$$\bar{u} = u^*\gamma^4, \quad \bar{v} = \gamma^4 v^* .$$

The normalization of the spinor field is such that:

$$\int d^3x u^*(x)u(x) = 1, \quad \int d^3x v^*(x)v(x) = 1$$

and also the $\zeta(t)$ are normalized.

If we now want to describe an initial state with τ_1 photons in the state 1, τ_2 in the state 2 ..., τ_α in the state α, we can write the permanent (from now on we shall use a short notation, neglecting all vector and spinor indices):

$$\varphi_I^{(b)} = \frac{1}{\sqrt{\tau_1!\tau_2!\dots\tau_\alpha!}\;a!} \begin{bmatrix} \zeta(1) & \zeta(2) & \dots & \zeta(a) \\ t_1 & t_2 & \dots & t_a \end{bmatrix}$$

$$= \varphi(t_1 t_2 \dots t_a); \quad [\zeta^{(r)} t_h] = \zeta^{(r)}(t_h) \,. \tag{2.16}$$

For final photons we have ($\tau_{1'}$ photons in the state 1′, etc.):

$$\varphi_F^{(b)} = \frac{1}{\sqrt{\tau_{1'}!\tau_{2'}!\dots\tau_{\beta'}!}\;b!} \begin{bmatrix} \zeta(1') & \zeta(2') & \dots & \zeta(b') \\ t_{a+1} & t_{a+2} & & t_{P_o} \end{bmatrix}$$

$$= \varphi(t_{a+1}, t_{a+2} \dots t_{P_0}) \,. \tag{2.17}$$

Proceeding in the same way, for the initial and final electrons and positrons we get the determinants:

$$\psi_I^{el}(y_1 y_2 \dots y_n) = \frac{1}{\sqrt{n!}} \begin{pmatrix} u^{(1)} & u^{(2)} & \dots & u^{(n)} \\ y_1 & y_2 & \dots & y_n \end{pmatrix};$$

$$(u^{(k)} y_h) = u^{(k)}(y_h) \,, \tag{2.18}$$

$$\psi_I^{pos}(x_1 x_2 \dots x_m) = \frac{1}{\sqrt{m!}} \begin{pmatrix} v^{(1)} & v^{(2)} & \dots & v^{(m)} \\ x_1 & x_2 & \dots & x_m \end{pmatrix};$$

$$(v^{(k)} x_h) = v^{(k)}(x_h) \,, \tag{2.19}$$

$$\psi_F^{el}(x_{m+1}x_{m+2}\dots x_{N_0}) = \frac{1}{\sqrt{p!}}\begin{pmatrix} u^{(1')} & u^{(2')} & \dots & u^{(p')} \\ x_{m+1} & x_{m+2} & \dots & x_{N_0}\end{pmatrix}, \tag{2.20}$$

$$\psi_F^{pos}(y_{n+1},y_{n+2}\dots y_{N_0}) = \frac{1}{\sqrt{q!}}\begin{pmatrix} v^{(1')} & v^{(2')} & \dots & v^{(q')} \\ y_{n+1} & y_{n+2} & \dots & y_{N_0}\end{pmatrix}. \tag{2.21}$$

(The adjoints $\overline{\Psi}$ are defined by writing in the Slater determinants the adjoints $\overline{u}$, $\overline{v}$ of the single-particle wave functions u, v.)

We can finally give the result in the desired form. Define a propagator as the function:

$$K\left(\begin{matrix} x_1x_2\dots x_{N_0} \\ y_1y_2\dots y_{N_o}\end{matrix}\middle|\, t_1t_2\dots t_{P_o}\right) =$$

$$\sum_{N(P_0)} \frac{\lambda^N}{N!}\int d1 \sum_1 \gamma^1 \dots \int dN \sum_N \gamma^N \begin{pmatrix} x_1x_2 & \dots x_{N_0} & 1\ 2 \dots N \\ y_1y_2 & \dots y_{N_0} & 1\ 2 \dots N\end{pmatrix}.$$

$$\cdot\,[t_1t_2\dots t_{P_0}\quad 1\ 2\dots N], \tag{2.22}$$

where $\sum_{N(P_0)}$ means summation over all the values of N having the same parity as P_0.

Call Φ_I the product of (2.16), (2.18) and (2.19), Φ_F the product of (2.17), (2.20) and (2.21) Then we can write the matrix elements $M_{FI}(\tau_1\,\tau_0)$ (τ_1 and τ_0 are times) as:

$$M_{FI}(\tau_1,\tau_0) = C_{FI}\,\widehat{\int}\Phi_F\, K\left(\begin{matrix} x_1x_2\dots x_{N_0} \\ y_1y_2\dots y_N\end{matrix}\middle|\, t_1t_2\dots t_{P_o}\right)\Phi_I \tag{2.23}$$

where C_{FI} is simply a numerical coefficient and the integration is performed over all the variables $x_1 \dots x_{N_0}$, $y_1 \dots y_{N_0}$, $t_1 \dots t_{P_0}$. In the symbol $\hat{\int}$ is included a time average, which we introduce in order to avoid the adiabatic switching on and off of the charge. For instance if w is the wave function of an initial particle:

$$\hat{\int} \dots \; w(x) = \lim_{T \to \infty} \frac{1}{\tau_0 + T} \int_{-T}^{\tau_0} dx_0 \int d^3x \dots w(x).$$

Formula (2.22) gives all the contributions of all Feynmann's graphs with N_0 external fermion lines and P_0 external boson lines.

2.8 Two Particular Cases

We treat briefly the cases in which only the boson or only the fermion field is quantized.

(i) Boson field quantized. In this case the bosons are in interaction with a classical current $j_\mu(x)$; the perturbation expansion can be summed exactly. No determinants appear in the expansion but only hafnians. For instance the vacuum-vacuum probability amplitude is given by

$$M_{00} = \sum_{n=0}^{\infty} \frac{(i)^{2n}}{(2n)!} \int d1 \sum \dots \int d(2n) \sum j(1) j(2) \dots$$

$$\dots j(2n) \left[1 \; 2 \; \dots \; 2n\right] =$$

$$= \sum_{n=0}^{\infty} (-1)^n \frac{1}{2^n n!} \left[\iint d1 \; d2 \; j(1) \left[1 \; 2\right] j(2)\right]^n =$$

$$= \exp\left[-\frac{1}{4} \iint d1 \; d2 \; j(1) \; D^F(1-2) j(2)\right]$$

which is Glauber's result.

(ii) Fermion field quantized. In this case no hafnians appear in the expansion but only determinants. For instance the scattering of one electron by an external electromagnetic field $A_\mu(x)$ can be described by a kernel $\widetilde{K}\binom{x}{y}$ whose perturbative expansion is:

$$\widehat{K}\binom{x}{y} = \sum_{N=0}^{\infty} \frac{\lambda^N}{N!}\int d1 \sum_1 \ldots \int dN \sum_N \gamma A_\mu(1) \ldots \gamma^\mu A_\mu(N) \begin{pmatrix} 1 & 2 & \ldots & N_x \\ 1 & 2 & \ldots & N_y \end{pmatrix} \tag{2.25}$$

This is just the Fredholm expansion and it is well known that:

$$\widehat{K}\binom{x}{y} = (xy)\,\widetilde{K}_0 - \lambda \int (x\,1)\gamma^\mu A_\mu(1)\,\widehat{K}\binom{1}{y} \tag{2.26}$$

where

$$\widetilde{K}_0 = \sum_{N=0}^{\infty} \frac{\lambda^N}{N!}\int d1 \sum_1 \ldots \int dN \sum_N \gamma^\mu A_\mu(1) \ldots \gamma^\mu A_\mu(N) \cdot \begin{pmatrix} 1 & 2 & \ldots & N \\ 1 & 2 & \ldots & N \end{pmatrix}.$$

(The usual notations are $\widetilde{K}\binom{x}{y}/\widehat{K}_0 = K_+^A(xy)$.

3. BRANCHING EQUATIONS FOR PROPAGATORS

Equations connecting propagators can be, and have been, derived in a variety of ways, from functional differentiation techniques to straight forward derivation from their perturbative expansions. We recall the standard definition of a Green function, viz.,

$$G_{N_0,P_0} = G\left(\begin{matrix} x_1, \ldots x_{N_0} \\ y_1, \ldots y_{N_o} \end{matrix} \middle| \; t_1, \ldots t_{P_o} \right)$$

$$= \frac{K_{N_0,P_o}\left(\begin{matrix} x_1, \ldots, x_{N_0} \\ y_1, \ldots, y_{N_o} \end{matrix} \middle| t_1, \ldots t_{P_o} \right)}{K_{00}}$$

We shall find equations of two types: type I contains only propagators K_{N_0,P_0} but type II also contain first order derivatives of K_{N_0,P_0} with respect to the masses and charges of the theory. Therefore, only the type I equations can be written in a similar form for the Green's functions G_{N_0,P_0}.

We list them briefly here. We note first that in the formal theory the branching equations among propagators can be written either as recurring hyperbolic differential equations with specific boundary conditions (in our case, causality requirements), or, equivalently, as recurring integral equations, the solutions of which (if any) satisfy both the differential equations and the boundary conditions. It is, however, in the indiscriminate transition from the first to the second way of formulating the problem that most of the troubles due to the so-called ultraviolet divergences are originated; as is well known, indeed, this formal transition is entirely meaningless already in the case of ordinary hyperbolic equations.

The case of a quantized electron field in an external E.M. field - in which the first traces of the kernel of the corresponding Fredholm equation are infinite, so that one must add special prescriptions, which were called also renormalization before their trivial origin was made evident[8)] - is also instructive. The correct transition from differential to integral equations obtains only if special cautions are taken, such as using a redefined concept of integral, as first introduced by

Hadamard[6].

One then has the choice, to start directly from the differential formulation of the problem and devise correct methods for its solutions, or to reconsider the integral branching equations so as to remove all ambiguities and troubles from their definition. We follow the second approach, which is more convenient and, of course, entirely equivalent to the first one.

In electroydnamics the branching equations among the propagators are:

Type I

$$K\begin{pmatrix} x_1 \ldots x_{N_0} \\ y_1 \ldots y_{N_0} \end{pmatrix} t_1 \ldots t_{P_0} \Bigg) =$$

$$= \sum_{h=2}^{P_0} [t_1 t_h] \, K \begin{pmatrix} x_1 \ldots x_{N_0} \\ y_1 \ldots y_{N_0} \end{pmatrix} t_2 \ldots t_{h-1} t_{h+1} \ldots t_{P_0} \Bigg)$$

$$+\lambda \int d\xi \sum \gamma^{\xi} [t_1 \xi] K \begin{pmatrix} x_1 \ldots x_{N_0} \\ y_i \ldots y_{N_0} \end{pmatrix} t_2 \ldots t_{P_0} \Bigg), \tag{3.1}$$

$$K_{N_0 P_0} = \sum_{h=1}^{N_0} (-1)^{h-1} (x_1 y_h) K \begin{pmatrix} x_2 \ldots x_{N_0} \\ y_1 \ldots y_{h-1} \, y_{h+1} \ldots y_{N_0} \end{pmatrix} t_1, \ldots t_{P_0} \Bigg)$$

$$-\lambda \int d\xi \sum \gamma^{\xi} (x_1 \, \xi) K \begin{pmatrix} x_2 \ldots x_{N_0} \\ y_1 y_2 \ldots y_{N_0} \end{pmatrix} t_1 \ldots t_{P_0} \Bigg), \tag{3.2}$$

$$K\begin{pmatrix} x_1 \ldots x_{N_0} \\ y_1 \ldots y_{N_0} \end{pmatrix} = \sum_{h=1}^{N_0} (-1)^{h-1} (x_1 y_h) K \begin{pmatrix} x_2 \ldots x_{N_0} \\ y_1 \ldots y_{h-1}\, y_h \ldots y_{N_0} \end{pmatrix}$$

$$- \lambda^2 \int d\xi_1 \int d\xi_2 \sum \gamma^{\xi_1} \gamma^{\xi_2} (x_1 \xi_1) [\xi_1 \xi_2]$$

$$K \begin{pmatrix} \xi_1 x_2 \ldots x_{N_0}\, \xi_2 \\ y_1 y_2 \ldots y_{N_0}\, \xi_2 \end{pmatrix} \quad . \tag{3.3}$$

<u>Type II</u>

$$\frac{\partial K_{N_0 P_0}}{\partial \lambda} = \int d\xi \sum \gamma^{\xi} K \left(\begin{matrix} x_1 \ldots x_{N_0} \xi \\ y_1 \ldots y_{N_0} \xi \end{matrix} \middle| \xi\, t_1 \ldots t_{P_0} \right) , \tag{3.4}$$

$$\frac{\partial K_{N_0 P_0}}{\partial m_f} = - i \int d\xi \; K \left(\begin{matrix} \xi\, x_1 \ldots x_{N_0} \\ \xi\, y_1 \ldots y_{N_0} \end{matrix} \middle| t_1 \ldots t_{P_0} \right), \tag{3.5}$$

$$\frac{\partial K_{N_0 P_0}}{\partial (m_b^2)} = - \frac{i}{2} \int d\xi \; K \left(\begin{matrix} x_1 \ldots x_{N_0} \\ y_1 \ldots y_{N_0} \end{matrix} \middle| \overset{o}{\xi}\, \overset{o}{\xi}\, t_1 \ldots t_{P_0} \right) . \tag{3.6}$$

The superfix ^{o} above a variable means that in expanding

$$[x_h\, y_k] = 0, \quad [x_h^o\, x_k] = [x_h\, x_k^o] = [x_h\, x_k]$$

Type II equations are better written <u>without</u>

this notation (which is inconvenient when iterations are performed); for this it suffices to define, (as for example in the neutral scalar meson $g\varphi^4$ case which we now give) the branching equations starting from:

$$K_{2n}(x_1 x_2 \dots x_{2n}) =$$
$$= \sum_{N=0}^{2n} \frac{\lambda^N}{N!} \int d\xi_1 \, .. \int d\xi_N \, [x_1 \dots x_{2n} \xi_1 \xi_1 \xi_1 \xi_1 \dots \xi_N \xi_N \xi_N \xi_N]. \tag{3.7}$$

and the modified propagators,

$$\widehat{K}_{2n} = \exp\left[-\frac{i}{2} \int d\xi \int^{\mu} [\xi \xi]_{\mu'} \, d\mu' \right] K_{2n}, \tag{3.8}$$

where $\lambda = -ig$, and $\mu = m^2$. is the volume of integration.

We then have

(i) equations of type I

$$\widetilde{K}_{2n}(x_1 \dots x_{2n}) = \sum_{h=2}^{2n} [x_1 x_h] \, \widehat{K}_{2n-2}(x_2 \dots x_{h-1} \, x_{h+1} \dots x_{2n})$$
$$+ 4\lambda \int d\xi \, [x, \xi] \widehat{K}_{2n+2}(x_2 \dots x_{2n} \xi \xi \xi), \tag{3.9}$$

(ii) equations of type II

$$\partial_\lambda \widetilde{K}_{2n}(x_1 \dots x_{2n}) = \int \widehat{K}_{2n+4}(x_1 \dots x_{2n} \xi \xi \xi \xi) d\xi \, , \tag{3.10}$$

$$2i \, \partial_\mu \widehat{K}_{2n}(x_1 \dots x_{2n}) = \int \widehat{K}_{2n+2}(x_1 \dots x_{2n} \xi \xi) d\xi \tag{3.11}$$

If we define the ratios

$$G_{2n} = \frac{K_{2n}}{K_o} = \frac{\widetilde{K}_{2n}}{\widehat{K}_o}$$

(Green's functions), then only the equations (3.9) retain their linear form:

$$G_{2n}(x_1 \ldots x_{2n}) = \sum_{h=2}^{2n} [x_1\ x_h]\ G_{2n-2}(x_2 \ldots x_{h-1}\ x_{h+1} \ldots x_{2n}) +$$

$$+ 4\lambda \int [x_1 \xi]\ G_{2n+2}(x_2 \ldots x_{2n} \xi\xi\xi) d\xi$$

with $G_o = 1$.

4. NUMERICAL MODELS

4.1 The Spurion Model

The simplest model, which has nothing to do with physics (it describes if one likes, a "spurion field") is obtained if we take

$$[x\ y] = \text{const.} = f(\mu) = \frac{1}{\Omega(a-\mu)}$$

in equations (3.9) to (3.11); (a is a constant and the form of $f(\mu)$ is required for consistency with (3.11)). It is easily seen from (3.9) to (3.11) that the relation

$$\partial_\lambda \int d\xi_1 \int d\xi_2\ \hat{K}_{2n}(x_1 x_2 \ldots x_{2n-4} \xi_1 \xi_1 \xi_2 \xi_2)$$

$$+ 4\partial_\mu^2 \int d\xi\ \hat{K}_{2n}(x_1 \ldots x_{2n-4}\ \xi\xi\xi\xi) = 0 \qquad (4.1)$$

holds. It is also easy to see directly from (3.9) to (3.11) that, for our model,

$$\lambda^2 \frac{d^2}{d\lambda^2} \tilde{K}^M_{2n} + \left(\frac{4n^2+2n}{2}\right)\lambda - \frac{1}{16\Omega f^2} \frac{d}{d\lambda} \tilde{K}^M_{2n} + \frac{4n^2+8n+3}{16} \tilde{K}^M_{2n} = 0$$

(4.2)

This equation is formally solved by the obviously divergent expansion

$$\sum_{N=0}^{\infty} \frac{(4N+2n-1)!!}{N!} \Omega^N f^{2N+n} \lambda^N = \sum_{N=0}^{\infty} \frac{a_N}{N!} \lambda^N = K_{2n}^M \tag{4.3}$$

The interest of such a model lies in the fact that, on the one hand, we can solve exactly (4.2) with the wanted initial conditions, which amounts to the same as solving the equations obtained for the model from equations (3.9) to (3.11) and on the other hand we can pose and solve the problem, how can one use the coefficients a_N of the formal divergent expansion (4.3) to obtain again the same exact solution of (4.2).

The general integral of (4.2) is given by

$$\tilde{K}_{2n}^M = A_{2n} y^{(2n+1)/4} \Phi\left(\frac{2n+1}{4}, \frac{1}{2}, -\frac{y}{4}\right) + B_{2n} y^{(2n+3/4} \cdot \Phi\left(\frac{2n+3}{4}, \frac{3}{2}, -\frac{y}{4}\right), \tag{4.4}$$

where we have put

$$y = \frac{1}{4\Omega f^2 \lambda}$$

and Φ denotes the Kummer function.

Equation (4.2) has an irregular singular point at $\lambda = 0$; its integrals are therefore expected to present a singular behaviour at the same point. It is nevertheless possible to obtain asymptotic (and diverging) series which represent the functions (4.4) for $y \to \infty$ (i.e., $\lambda \to 0$). One must distinguish two cases ($(a)_n = \Gamma(a+n)/\Gamma(a)$):

a) Re $y > 0$

$$(\tilde{K}^M_{2n})^{\mathrm{Re}\, y>0}_{asy} = A_{2n} 2^{(2n+1)/2} \frac{\Gamma(\frac{1}{2})}{\Gamma(-(2n-1)/4)} \cdot$$

$$\cdot \sum_{N=0}^{\infty} \binom{-(2n+3)/4}{N} \frac{4^N((2n+1)/4)_N}{(-y)^N}$$

$$+ B_{2n} 2^{(2n+3)/2} \frac{\Gamma(\frac{3}{2})}{\Gamma(-(2n-3)/4)} \cdot$$

$$\cdot \sum_{N=0}^{\infty} \binom{-(2n+1)/4}{N} \frac{4^N((2n+3)/4)_N}{(-y)^N} ; \tag{4.5}$$

b) Re $y < 0$

$$(\tilde{K}^M_{2n})^{\mathrm{Re}\, y<0}_{asy} = A_{2n} \exp\left[-\frac{y}{4}\right] y^n \left(-\frac{1}{4}\right)^{(2n-1)/4} \frac{\Gamma(\frac{1}{2})}{\Gamma((2n+1)/4)} \cdot$$

$$\cdot \sum_{N=0}^{\infty} (-1)^N \binom{(2n-3)/4}{N} \frac{(-(2n-1)/4)_N 4^N}{(-y)^N}$$

$$+ B_{2n} \exp\left[-\frac{y}{4}\right] y^n \left(-\frac{1}{4}\right)^{(2n-3)/4} \frac{\Gamma(\frac{3}{2})}{\Gamma((2n-3)/4)} \cdot$$

$$\cdot \sum_{N=0}^{\infty} (-1)^N \binom{(2n-1)/4}{N} \frac{(-(2n-3)/4)_N 4^N}{(-y)^N} \tag{4.6}$$

We start, for the sake of simplicity, with n = 0.

Then (4.5) reduces to

$$(\hat{K}_0^M)_{asy}^{Re\ y>0} = \left(\frac{2^{\frac{1}{2}}\Gamma(\frac{1}{2})}{\sqrt{\pi}\ \Gamma(\frac{1}{4})} A_o + \frac{2^{\frac{1}{2}}\ \Gamma(\frac{1}{2})}{\sqrt{\pi}\ \Gamma(\frac{3}{4})} B_o \right) \cdot \sum_{N=0}^{\infty} \frac{\Gamma(\frac{1}{2})(4N-1)!!}{N!2^{2N}y^N}$$

$$= \sqrt{2}\ \Gamma(\tfrac{1}{2})\left(\frac{A_o}{\Gamma(\frac{1}{4})} + \frac{B_o}{\Gamma(\frac{3}{4})}\right) \sum_{N=0}^{\infty} a_N \frac{\lambda^N}{N!} \tag{4.7}$$

with the a_N defined in (4.3).

If therefore, we take in case a)

$$\sqrt{2}\ \Gamma(\tfrac{1}{2})\left(\frac{A_o}{\Gamma(\frac{1}{4})} + \frac{B_o}{\Gamma(\frac{3}{4})}\right) = \left(\frac{\sqrt{2}f}{i}\right)^{\frac{1}{2}} ,$$

we find that all the (infinitely many) integrals (4.4) of equation (4.2) which satisfy this condition admit the asymptotic expansion (4.3). This is also the general case, as long as Re $y>0$; for $n \neq 0$ it would be straightforward to verify that the same result holds, provided A_{2n} and B_{2n} are given (as is required by (3.10) and (3.11)) by

$$A_{2n} = (-1)^{(2n+1)/4}\ 2^{(2n-3)/2}\ f^n\Gamma\left(\frac{2n+1}{4}\right)\left[\frac{1+(-1)^n}{\Gamma(\frac{1}{4})} A_o + i\ \frac{1-(-1)^n}{\Gamma(\frac{3}{4})} B_o\right] \tag{4.9}$$

$$B_{2n} = (-1)^{(2n+3)/4}\ 2^{(2n-1)/2}\ f^n\ \Gamma\left(\frac{2n+3}{4}\right)\left[i\ \frac{1+(-1)^n}{\Gamma(\frac{3}{4})} B_o + \frac{1-(-1)^n}{\Gamma(\frac{1}{4})} A_o\right] , \tag{4.10}$$

where A_o and B_o satisfy the condition (4.8).

Let us now consider case b). If, for $\operatorname{Re} y < 0$, we take values of A_o and B_o which satisfy (4.8) the resulting asymptotic expansion does not coincide with the formal perturbative expansion, except for a special choice of values of A_{2n} and B_{2n}, which reduce (introducing the Tricomi functions Ψ) to

$$\tilde{K}^M_{2n} = \sqrt{-i\Omega}\,(2n-1)!!\,\left(\frac{ify}{2}\right)^{(2n+1)/4}\Psi\left(\frac{2n+1}{4}, \frac{1}{2}, -\frac{y}{4}\right); \tag{4.11}$$

these values are now uniquely determined as

$$A_{2n} = \frac{\Gamma(\frac{1}{2})}{\Gamma((2n+3)/4)}\,(2n-1)!!\,\sqrt{-i\Omega}\,\left(\frac{if}{2}\right)^{(2n+1)/2}, \tag{4.12}$$

$$B_{2n} = \frac{\Gamma(-\frac{1}{2})}{\Gamma((2n+1)/4)}\,(2n-1)!!\,\frac{\sqrt{i\Omega}}{2}\,\left(\frac{if}{2}\right)^{(2n+1)/2}. \tag{4.13}$$

In conclusion we find a completely different behaviour of the solutions of our model, according as we have $\operatorname{Re} y > 0$ or $\operatorname{Re} y < 0$ at the irregular singular point $y = \infty$ ($\lambda = 0$). If $\operatorname{Re} y > 0$, there are infinitely many solutions with the wanted asymptotic expansion (the difference between any two of them is proportional to the non-vanishing solution having vanishing asymptotic expansion at $\lambda = 0$); if $\operatorname{Re} y < 0$, there is only the solution (4.11) which has this behaviour and can be continued to all points with $\operatorname{Re} y > 0$, with the exception of the positive real semi-axis, which is a cut in the y-plane). If $\operatorname{Re} y = 0$, the asymptotic expansion is the sum of (4.5) and (4.6) so that the considerations made for $\mathrm{Re}\, y < 0$ cover this case too.

We can now study the following interesting problem, namely to use the knowledge that the divergent expansion (4.3) is a formal solution of equation (4.2) in order to determine a convergent solution of this equation. This task is particularly simple, due to the elementary character of the model. As is well known, whereas to any given function there corresponds a unique asymptotic expansion, the converse is false: the solution of this problem cannot be expected a priori to be unique in general: we know already that this is not the case in our model, which provides a sufficient counter-example also for the general regularized theory.

Equation (4.2) is of type

$$\lambda^2 u''(\lambda) + (a\lambda + b)u'(\lambda) + cu(\lambda) = 0 \qquad (4.14)$$

Our procedure is based upon the remark, the verification of which is trivial, that if

$$\sum_{N=0}^{\infty} A_N \lambda^N \qquad (4.15)$$

is a formal solution of (4.14) (or of any similar equation having as coefficients polynomials in λ), then also the expression which is obtained by substituting in (4.15) λ^N with a suitable linear integral transform or limit, and then exchanging the $\sum_N$ in (4.15) with the other infinite operation thus introduced, is a formal solution of (4.14). We may put thus

$$\lambda^N = \int_0^\infty \exp\left(-\frac{t}{\lambda^{1/k}}\right) \frac{t^{KN-1}}{\Gamma(K\,N)}\, dt \qquad (4.16)$$

or

$$\lambda^N = \lim_{\rho \to 1} (1-\rho)^{-N} \mathcal{D}^{+}_{-2N}\left(\frac{1}{\sqrt{(1-\rho)}}\right)$$

$$-\frac{3}{4} < \arg \frac{1}{\sqrt{(1-\rho)}} < \frac{3}{4}, \quad \rho \leq 1 \qquad (4.17)$$

(which comes from a Gauss transformation), where $\mathcal{D}_\nu^{+}$ is a parabolic cylinder function of order ν, and obtain again, after exchange of $\sum$ with $\int$ or lim, a formal solution of (4.14).

This offers the solution to our problem in all cases in which, starting from a formal divergent (and therefore meaningless) solution (4.15) of an equation of type (4.14)) we can find that by exchanging $\sum$ with $\int$ the resulting expression is instead convergent, and therefore yields, with all due rigor because of what we have just seen, a correct solution of (4.14).

In other words, the exchange of two infinite operations can be used with rigor and profit without any need of further proofs, under the conditions just stated, to obtain from the divergent expansion (4.15) a convergent correct solution. We may mention, in passing, that several divergences which are typical of field theory can be gotten rid of by means of similar simple devices, which have proved quite effective in various cases[10, 11].

We now proceed to apply this method to the formal solution (4.3) of the equation (4.2) which defines our model. We take, for the sake of simplicity, $n = 0$, (the same is true for all values of n). We obtain, using the definition of y:

$$(K_0^M)_{asy} = \sum_{N=0}^{\infty} \frac{(4N-1)!!}{N!} \frac{1}{(4y)^N} \quad . \tag{4,18}$$

To cover at the same time all cases $\mathrm{Re}\, y \gtreqless 0$ it is convenient to use the transformation

$$y^{-(N+\frac{1}{4})} = \lim_{\alpha \to 0} \int_0^{\infty} \exp[-(\sqrt{-y} + \alpha)t] \frac{t^{2N-\frac{1}{2}}}{\Gamma(2N+\frac{1}{2})} dt$$

with $\mathrm{Re}\, \alpha > 0$ (except in the limit); one has to take $\sqrt{-y}$ on the first branch if $\mathrm{Re}\, y \geqslant 0$, on the second if $\mathrm{Re}\, y < 0$.

Substituting (4.19) into (4.18) and exchanging $\sum$ with $\lim \int$ we obtain

$$(K_0^M)_{asy} = \frac{-(y)^{\frac{1}{4}}}{\sqrt{\pi}} \sum_{N=0}^{\infty} (-1)^N \frac{\Gamma(2N+\frac{1}{2})}{N!} \lim_{\alpha \to 0} \int_0^{\infty} \cdot$$

$$\exp[-(\sqrt{-y} + \alpha) t] \frac{t^{2N-1}}{\Gamma(2N+\frac{1}{2})} dt$$

$$= \frac{(-y)^{\frac{1}{4}}}{\sqrt{\pi}} \lim_{\alpha \to 0} \int_0^{\infty} \frac{\exp[-(\sqrt{y} + \alpha)\, t]}{\sqrt{t}} \exp[-t^2] dt$$

$$= \frac{(-y)^{\frac{1}{4}}}{\sqrt{2}} \left[\frac{\sqrt{\pi}}{\Gamma(\frac{3}{4})} \Phi(\tfrac{1}{4}, \tfrac{1}{2}, -\tfrac{y}{4}) + \frac{\sqrt{\pi}}{\Gamma(\frac{1}{4})} \sqrt{\frac{-y}{4}}\, \Phi(\tfrac{3}{4}, \tfrac{3}{2}, -\tfrac{y}{4}) \right]; \tag{4.20}$$

this is the value (4.11) of K_0^M as deduced from integration of (4.2) if y is not on the cut (which we may take on the real positive y semi-axis).

We find thus the correct solution when this is unique, one of the solutions when there is no uniqueness. The second sign of equality in (4.20) is of course to be understood as equating two formal solutions of (4.14); the convergence of the latter expression is a sufficient proof of existence; after what we have seen, the legitimization of this procedure lies in the form of equation (4.14) and no further demonstration is necessary in this, or any similar case.

As a final remark, we note that our finding a solution analytical in $1/\sqrt{g}$ is just what we would expect from the c-number theory.

One then has an equation

$$\Box\varphi - m^2\varphi - g\varphi^3 = 0 ,$$

which is solved on inspection in the limiting case $m = 0$, for $\varphi = \varphi(s)$, $s^2 = x^2 + y^2 + z^2 - c^2t^2$. In fact then this equation reduces to

$$s\frac{d^2}{ds^2}\varphi + \frac{d\varphi}{ds} - sg\varphi^3 = 0 ,$$

which has the solution

$$\varphi = \frac{1}{\sqrt{g}\sqrt{x^2 + y^2 + z^2 - c^2t^2}} .$$

The cubic power term causes the solution to be singular in g: it is dominant with respect to linear terms, and cannot be neglected even if g is very small. We find, thus, that the c-number theory gives essentially the same type of behaviour as our highly idealized numerical model, which yet retains all the features of Bose statistics and the ensuing algebraic complications.

4.2 Divergence of a Perturbative Expansion

The method followed here to obtain a correct convergent solution from the divergent expansion of the model is of quite wide generality, and is applicable with profit, e.g. to questions which arise when singular potentials are treated. It consists in systematizing procedures which are often used heuristically in mathematics, by turning to our advantage the fact that infinite operations cannot be generally exchanged, to use just such an exchange, provided it does not alter some formal properties of the expressions it acts upon, to obtain the convergent solution from a divergent one.

More precisely, given a set of equations, any series of functions (powers or not) which does satisfy term-wise those equations, can be in full right regarded as a *bona fide* solution, regardless of whether it converges or not; in the first case its sum defines a function and there is no further arguing, in the second it still is with all rigor a *formal solution* (which may serve the more limited purpose of providing an asymptotic evaluation of the true solution, if any). Since the fact that a series satisfies the given equations is quite independent from its convergence (at least if a finite number of terms are involved at each step of the verification, as is the case in the instances of interest to us), one may attempt transforming, with an integral transformation or some other infinite process, the given formal series into a new one, which is easily proved (once for all) to be a formal solution of the transformed equations; if the latter series happens to converge, and if the function to which it converges can be anti-transformed, the anti-transform thus obtained is clearly the wanted solution, which is thus obtained by exchanging an infinite sum with another infinite operation.

If the situation in actual field theories is anything like the one we find in our numerical model, then the study of their existence and uniqueness is a problem of <u>real variable</u>, and analytical functions are of little avail. It may be important, therefore, to recognize what properties might replace analyticity in such cases; (our finding that the regularized $g\phi^4$ theory leads to solutions, if any, of class 2, might deserve in this light deeper investigation[16]). This is a task for mathematicians, who have thus far dedicated but little attention to these problems. For convenience of the reader, and in the hope that this may stimulate further thought on the subject, we recall here briefly some basic notions about these more general classes of functions.

A typical property of analytical functions is that they are uniquely determined in the whole domain of definition if their values are known within an arbitrarily small region of that domain. They are not, however, the only functions thus characterized; the same requirement is satisfied by all, and only the functions which belong to the wider class of <u>quasi-analytical</u> functions, thoroughly studied by Carleman[6]. The uniqueness of the result of the continuation process depends on the maximum values assumed by the derivative of the function (supposed to be infinitely many times differentiable in the whole domain) within the domain of definition; if a function $f(x)$ in $\mathcal{D}$ is such that $|f^{(n)}(x)| < KA_n$ we may say that $f(x)$ is of class $\mathcal{A} \equiv (A_0, A_1, \ldots)$, $f(x)$ is analytical if $A_n = n!$; a sufficient condition for quasi-analyticity is that the series $\sum_{n=0}^{\infty} A_n^{-1/n}$ be divergent (Denjoy[7]); a condition which is also necessary is given by Carleman[6]. If, however, one just takes $A_n = \Gamma(1+(1+\varepsilon)n)$, which can be, for given n, as close as wanted to $n!$, it can be proved that $f(x)$ can be continued in infinitely many ways from the point $x = 0$ $(0 \in \mathcal{D})$.

If $A_n = \Gamma(1+\alpha n)$ then the function $f(x)$ can be described, more simply, as being of class α (Hadamard[8], Gevrey[9]). An important case is that of the functions of Hadamard's class 2: $A_n = (2n)!$ Such is, for instance, the function defined in the real interval $(0, a)$ by $\exp[-1/x]$, which can be continued in infinitely many ways on the negative semi-axis (e.g., with the vanishing function); this is the classical example of a function with vanishing asymptotic expansion at $x = 0$. The solutions of the heat equation are, in general, of class 1 (analytical) in the space variable and of class 2 in the time variable.

The solutions of the equations of our numerical model are also essentially of class 2, as will be easily surmised from the expansions reported later. The same will be seen to be true also for the solutions of the regularized $g\varphi^4$ theory.

We notice, finally, that the solutions of the numerical model are for generic values of $1/g$ (that is for $1/g \neq 0$, which is a branch point, and $1/g \neq \infty$, which is an essential singularity and a branch point) perfectly good holomorphic functions of $1/g$, in a cut plane. Most of our troubles derive therefore from our having assumed as expansionparameter g instead of $1/\sqrt{g}$ (the square root and an outside factor $g^{(2n+1)/4}$ remove the branch point): this indicates the usefulness of solvable numerical models like the present one to indicate possibly "good" expansion parameters, if the ones used in the perturbative expansion turn out to be "bad".

- * - * -

Another model of interest to us was considered by F. Guerra and M. Marinaro[3]; it consists of a spinor field interacting with a spurion field with scalar coupling. It will be more appropriate to discuss it after something has been said about renormalization because it shows how seriously renormalization can affect the analytic behaviour of a solution.

In these and in all other cases considered by us so far, we have always found that explicit solutions, whenever they can be obtained, are, when considered as functions of the coupling constant, of the class 2 type of Hadamard.

5. FINITE PART INTEGRALS

5.1 An example

Before discussing F.P.I's and renormalization, it will be instructive to consider a most simple example, namely the Hilbert equation; this exhibits the basic techniques, to be used later in general, with a minimum of labour. The symbol used for the F.P.I. is $\rceil\!\!\int$: it is an operation acting on distributions or generalized functions, as will soon be specified. In this example

$$\rceil\!\!\int \equiv \quad \text{Cauchy P.V.} \int d\xi \ .$$

Our example is of interest because it shows, in a very simple way, how changing an ordinary Riemann integral - which is, under the circumstance, meaningless, just as in the formal branching equations of field theory - into a P.V. integral gives a well defined meaning to the equation and introduces "wave function" and "charge" renormalization.

Consider the equation

$$\varphi(x) + \lambda \int \frac{\varphi(\xi)}{x-\xi}\, d\xi = f(x) \ , \qquad (5.1)$$

where the integral ranges over, say a finite interval. Equation (5.1) is meaningless as it stands: all traces of $1/(x-\xi)$ are divergent.

It is known that an equation like (5.1), if written with P.V. integrals, is correctly stated and has well-defined solutions. Such is:

$$\bar{\varphi}(x) + \lambda \rceil\!\!\int \frac{\varphi(\xi)}{x-\xi}\, d\xi = f(x) \ . \qquad (5.2)$$

The question is, whether, and how, one can compensate for the "mistake" originally made by writing (5.1) with an ordinary integral, so as to change it into the form (5.2). We have, that is, to find the relation between $\overline{\int}$ and $\int$. This is very simple here: just remember that

$$\text{P.V.} \int \frac{\varphi(\xi)}{x-\xi}\, d\xi = \lim\Big(\int^{x-\epsilon'} + \int_{x+\epsilon''}\Big) \frac{\varphi(\xi)}{x-\xi}\, d\xi \,,$$

with the restriction $\epsilon' = \epsilon'' = \epsilon$, when $\epsilon \to 0$.

But the requirement $\epsilon' = \epsilon''$ is imposed only because one wishes to destroy the term

$$\lg\left|\frac{\epsilon'}{e''}\right| : \varphi(x) \,.$$

The same result can be obtained if that term is suppressed, rather than with the additional requirement $\epsilon' = \epsilon''$, by simply subtracting it, when of course ϵ', ϵ'' are left arbitrary but dealt with consistently; that is, by writing:

$$\int \frac{\varphi(\xi)}{x-\xi}\, d\xi = \overline{\int} d\xi \, \frac{(\xi)}{x-\xi} + d.\varphi(x) \,, \qquad (5.3)$$

where we put, for short, $\lg|\epsilon''/e'| = d$ and leave it understood that $\overline{\int}$ is calculated for the same (arbitrarily vanishing) ϵ', ϵ'' that came into the definition of d. The "divergent" - or, better, as we see from this example, "ambiguous" - part is given by

$$D^{\xi} \frac{\varphi(\xi)}{x-\xi} = d.\varphi(x) \,.$$

Substitution of (5.3) into (5.1) gives:

$$(1 + \lambda d)\varphi(x) + \lambda \overline{\int} \frac{\varphi(\xi)}{x-\xi}\, d\xi = f(x) \qquad (5.4)$$

which reduces to (5.2) if one takes:

$$\varphi(x) = \frac{\bar{\varphi}(x)}{1 + \lambda d}, \qquad (5.5)$$

$$\bar{\lambda} = \frac{\lambda}{1 + \lambda d}. \qquad (5.6)$$

(5.5) is a "wave function renormalization" and (5.6) a "charge renormalization", in the familiar language of field theory. This example shows - and we shall prove in the next work that this is the case for all renormalizable theories - that kf the equation is written in the "correct" form (5.2) to begin with, there is no need of (5.5) and (5.6) which serve only to compensate the faulty use of ordinary integrals in (5.1). A few more remarks may be in order. We note that in (5.2) is unknown a priori - experiment will decide on its numerical value. Thus, we may decide to use, instead of (5.3) the "subtractive" prescription (finite integration domain):

$$\int' \frac{\varphi(\xi)}{x - \xi} d\xi = \int d\xi \frac{\varphi(\xi) - (x)}{x - \xi};$$

$$D \frac{\varphi(\xi)}{x - \xi} = \varphi(x) \int \frac{d\xi}{x - \xi} = \delta \cdot \varphi(x);$$

we obtain a result which differs from (5.3) by a finite contribution $(\delta - d)$ to the "divergent part". Use of $\int'$ instead of $\int$ leads to an equation similar to (5.2), which reduces to (5.2) through a finite renormalization of type (5.5) and (5.6). But if the parameter $\bar{\lambda}$ is to be deduced from experiment, this will make no difference: no matter what rule is chosen to evaluate $\int$, all numerical predictions of the theory remain the same, provided $\bar{\lambda}$ is given the appropriate value each time.

We had here a logarithmic divergence: had it been quadratic (higher ones do not occur in equations of this type in renormalizable field theories), we would have found - as we shall see very soon - also a "mass renormalization", which is handled with the same ease.

5.2 Regularization and Renormalization

A thorough treatment of a specific field theory - for example, the standard $g\varphi^4$ coupling for neutral spinless mesons - requires too much formal labour; a full account of all details can be found in references 3) and 4); it will suffice here to indicate the main features of our method.

The formal theory, (be it written in the form of equations connecting Green's functions of various orders or their mass and charge derivatives, or of formal perturbative expansions) is clearly meaningless unless convenient regularizations are introduced to make all integrals finite and mutually commuting. Our treatment of renormalization purports to retain, throughout the renormalization procedure, the properties of finiteness and commutability of the multiple integrals of regularized theories; specifically, we make the following requirements.

(i) Renormalization should leave all Green's functions invariant, rearranging them internally so as to render them finite after the parameters (masses and charges) of the theory have incorporated all troublesome ambiguous or divergent contributions. This is Dyson's original criterion[12], failing which one might certainly have finite (regularized) theories, but yet not enough knowledge to assess their physical meaning without additional investigations.

(ii) Renormalization of the field equations should be wholly independent of the formal techniques used for their solution or approximation; any iteration procedure performed among them should always yield consistent results; in particular, the formal perturbative expansions (which are presumably divergent) of the renormalized equations should also come out correctly renormalized to each power, since they are obtained with an infinite number of iterations from the field equations.

(iii) The formal structure of the renormalized equations should be as simple as possible; combinatorics should take care automatically of all questions which are often studied with painstaking graph-by-graph computations (as was the case in the formal theory with our algebraic techniques, which extend the Fredholm formal treatment to coupled equations).

Since, in the actual case, one is faced with multiple integrations involving products of tempered distributions, the requirements set above are satisfied if the following logical steps can be accomplished.

a) Work first in the subspace S_o of $S(R^{4n})$ of test functions $\varphi_o(x_1 \ldots x_n)$ that vanish whenever any number of variables become coincident, with zeros of orders sufficiently high to warrant that the given product of tempered distributions be a linear continuous functional on S_o.

b) Extend then, with some well-specified procedure, all such functionals on S_o to functionals which are defined (and finite!) on the whole space $S(R^{4n})$ of test functions; the procedure adopted must be such (this requirement makes, in fact, the major difference between our treatment and those of other authors) as to preserve the same formal properties which are familiar with ordinary commuting multiple integrals of regular functions, for which the final value (i) is independent of the order of integrations and (ii) is the same whether all n integrals are performed together as one single 4n-dimensional integral, or as a sequence of successive 4-dimensional integrals.

It is then clear that this method of operating with distributions can still be denoted with some sort of symbol which has the same formal commutation properties as ordinary integrals: we may write, thus, $\rfloor dx_1 \ldots \rfloor dx_n$ and call $\rfloor$ a "finite-part integral", or F.P.I. (it is, actually, an extension to distributions of Hadamard's partie-finie integral[8], as we shall discuss later),

One procedure which accomplishes the wanted extension of functionals on S_o to functionals on $S(R^{4n})$ is discussed in the second paper of reference 3) and provved there to satisfy all the requirements; we shall briefly state it in the next section.

The whole program described thus far can then be accomplished. This extension, which ought to be regarded, to start with, only as a regularization,, will be straightforwardly found to be in fact a renormalization by using the formalism developed in our previous works[15]. The equations which connect renormalized Green's functions are entirely analogous to the formal, unrenormalized ones; this turns out to be quite advantageous in many ways, e.g. because it permits the demonstration of renormalizability of a field theory by analysing only the ultraviolet divergences which arise from confluence in a single point, so that all complicated overlap problems are automatically taken care of and by-stepped in the discussion. Renormalization thus becomes independent of perturbative techniques, or of any other method which may be used to solve or approximate the equations for Green's functions: provided the F.P.I. rule is specified, one may proceed without further worries about renormalization to handle them as best as one can. We refer to a previous work[13] for the proof that this method permits, also, starting from a Lagrangian or Hamiltonian, to recognize without explicit computations which terms must be added to it, if any, for the corresponding theory to be renormalizable.

The present work handles the transition between ordinary and F.P. integrals, and back: infinites, although correctly hidden, loom in the background nevertheless. A study of the infinitesimal renormalization group permits the rigorous proof that identical results hold also for the transition between <u>any</u> <u>two</u> different F.P. integration prescriptions[5]; infinite quantities are thus replaced by finite ones,(depending of course upon the prescription used), which may be computed in a well-defined manner. The ordinary integral thus becomes only a

<u>singular</u> case in a family of well-behaved F.P. integrals. A work on this subject[5)], shows that, starting from infinitesimal transformations, one can define a finite renormalization group which is quite similar to the one defined by Bogoliubov[14)]; the main difference is that, while Bogoliubov's treatment starts with particles having dressed, physical masses, our transformations start from bare masses, which are changed by renormalizations, as are the coupling constants and Green's functions.

5.3 Axioms for F.P.I's

Historically, this denomination was introduced by Hadamard[8)] (<u>partie finie</u>) to denote an operation which, by generalizing the Cauchy principal-value integral, permitted one to solve with the standard formal apparatus of the Green and Stokes relations some hyperbolic differential equations, without incurring the infinities which do notoriously arise if Riemann-Lebesgue integrals are, incorrectly, used instead. Various extensions of this concept were made later, which proved adequate to deal with problems involving <u>single</u> hyperbolic equations; in the meantime, theories of distributions or of generalized functions have grown into powerful tools, which have played a dominant role in most of the more recent works.

In the wider context of distribution theory the F.P. integral, as defined by Hadamard and the other authors mentioned, appears as a generalization of the ordinary concept of integral, whereby a distribution can be associated to a non-integrable function as follows. Let $f(x)$ be a function having a non-integrable singularity at $x = 0$, S a function space; then a F.P. integral can be regarded as an operation which engenders a correspondence between complex numbers C and functions $\varphi \in S$

$$f(x): S \rightarrow C,$$

$$\varphi(x) \rightarrow \langle f(x), \varphi(x)\rangle = \int f(x)\varphi(x)dx \quad (5.6)$$

subject to the requirement that, if the test functions φ are restricted to those φ_0, of a subspace $S_0 \subset S$, which vanish sufficiently fast at $x = 0$, it must yield

$$\int f(x)\varphi_0(x)dx = \int f(x)\varphi_0(x)dx . \quad (5.7)$$

Starting with $f(x)$, the ordinary integral defines thus a distribution on S_0, the F.P. integral a distribution on S; the latter is obtained from the former by means of a procedure of extension, the specification of which constitutes the "F.P. integration rule". This procedure is by no means unique and changes of it produce terms proportional to $\delta(x)$ and its derivatives. Thus, for instance, one may have, as we have seen:

$$\int f(x)\varphi(x)dx = \text{Cauchy P.V.} \int f(x)\varphi(x)dx$$

when this exists;

$$\int f(x)\varphi(x)dx = (-1)^n \int F(x)\varphi^{(n)}(x)dx ,$$

where

$$F(x) = \int^x dx_1 \int^{x_1} dx_2 \ldots \int^{x_{n-1}} dx_n f(x_n)$$

is integrable; or, by resorting to one among the possible procedures based upon analytic continuation,

$$\int f(x)\varphi(x)dx = \lim_{\lambda\to 0} \text{R.P.A.C.} \int f(x)x^{\lambda}\varphi(x)dx ,$$

where R.P.A.C. means "regular part of the analytical continuation" of the complex parameter λ. Examples can be multiplied at will; the latter mentioned above can be immediately generalized to n-dimensional integrations.

It is to be emphasized that F.P. integrals are symbols denoting operations which are in general (although not necessarily) more elaborate than ordinary integrations (such as performing an integral and afterwards a limiting process); the nature of these operations will depend heavily on the class of objects to be dealt with. Their use is very convenient in cases such as we meet in field theory, where their commutability and the legitimacy of performing the operations indicated by them, simultaneously or one at a time, (these properties are demonstrated explicitly to hold in reference (3) are essential in order to render all combinatorial manipulations with renormalized theories identical with the corresponding formal ones of unrenormalized theories; it is this feature that permits us to avoid in our formalism the customary detailed analysis of graphs of all sorts.

The prescriptions which must be assigned in order to define a F.P.I. adequate to deal with the situations which arise in quantum field theory are rather more elaborate than in the cases recalled above, which were treated by Hadamard and others with the more restricted scope of handling only singular functions. We have to consider in field theory products

$$T = \prod_{ij} D^c(x_i - x_j) \qquad (5.8)$$

of tempered distributions: our problem is therefore to find (at least) an extension of the linear continuous functionals defined thereby on $S_0(R^{4n})$ to functionals defined on $S(R^{4n})$; our extension must retain all the formal advantages mentioned above. The way we follow to discuss this problem is to state first all the wanted formal requirements as mathematical axioms, which must be satisfied by any "F.P. integration rule" to be acceptable in our treatment of renormalization; to exhibit then explicitly one such rule, and to prove, finally, that it does indeed satisfy the stated axioms (the last proof is given in reference (3).

The main feature of F.P. integrals as tools to compute physical quantities is that they are not, in general, uniquely defined. Their use is thus legitimate only if this ambiguity is physically irrelevant: the demand that it be so leads us immediately to recognize that this requisite is automatically satisfied only by some special field theories, which turn out to be exactly those which are renormalizable according to the standard Dyson classification. Renormalizability becomes thus, in this formalism, a simple requirement of mathematical consistency. By the same token one can recognize on inspection which terms must be added to the Lagrangian to make consistent, or renormalizable, a theory which is not such to start with[12].

Single and multiple F.P. integrals are defined through extensions of T from $S_0(R^{4n})$ to $S(R^{4n})$;

the "assumptions" made in the context must be proved explicitly for any given F.P. integration rule, equivalently, and more conveniently, they can be formalized in the following axioms, which any F.P.I. rule is required to satisfy.

a) Linearity and continuity (this entails in particular additivity with respect to integration domains).

b) Permanence

$$\int dx_1 \int_{x_1} dx_2 \dots \int_{x_1 \dots x_{n-1}} dx_n \, T(x_1 \dots x_n) \varphi(x_1 \dots x_n)$$

$$= \lim_{\varepsilon \to 0} \int dx_1 \dots \int dx_n T_\varepsilon (x_1 \dots x_n) \varphi(x_1 \dots x_n)$$

if the latter exists and is finite.

c) Generalized Fubini theorem

Full commutability of the F.P.I's must be assured.

The requirement that axiom (c) holds marks the main difference between other authors' treatments of renormalization and ours; this property permits us to eliminate divergences and ambiguities arising from confluences of any order by an iterative procedure which acts upon one variable at a time.

In addition, we require that two more axioms be satisfied; the first is necessary especially to handle branching equations of the 2nd type, the second only if the theory is relativistic, as in the present case.

d) <u>Commutability of F.P.I's with parameter derivatives</u>

If the test functions φ depend on a parameter α and both φ and $\partial\varphi/\partial x$ belong to the same functional space, it is required that:

$$\frac{\partial}{\partial\alpha} \int_{(x_1 \ldots x_n)} dx_1 \ldots dx_n \; T(x_1 \ldots x_n)\varphi(x_1 \ldots x_n, \;)$$

$$= \int_{(x_1 \ldots x_n)} dx_1 \ldots dx_n \; T(x_1 \ldots x_n)\frac{\partial\varphi}{\partial\alpha} \; .$$

e) <u>Relativistic invariance</u>

The relativistic invariance of the formal theory must be retained throughout F.P.I. renormalization.

The fulfilment of our program demands, as we have stated before, that the class of F.P. integration rules satisfying the axioms just stated,(whose existence was thus far only hypothesized) be shown to be non-empty. To do so it suffices to exhibit one specific rule and prove that it satisfies all the axioms; it becomes then a trivial matter to verify that the class contains in fact infinitely many equally satisfactory rules and that transition from one to the other is accomplished by means of a "finite renormalization". We shall report here only one specific prescription for F.P.I. and refer the reader to reference (3) for the proof that it does indeed satisfy all the axioms. Such a prescription is based on analytic continuation.

Let

$$\sigma_\varepsilon(x, \lambda) = (x^2 - i\varepsilon)^\lambda,$$

$$x^2 = g_{\mu\nu} x^\mu x^\nu ,$$

λ complex, be nuclei of auxiliary distributions. which we introduce to get

$$T_{(n,\ n-1)}(x_1 \ldots x_n) = \mathcal{O}(\lambda) T(x_1 \ldots x_n) \sigma(x_n - x_{n-1}, \lambda),$$

which is symbolic shorthand to denote that $T_{(n,n-1)}$ is a functional on $S_0^{(n,n-1)}$ defined by

$$\langle T_{(n,n-1)}(x_1 \ldots x_n), \psi(x_1 \ldots x_n) \rangle =$$

$$\mathcal{O}(\lambda) \lim_{\epsilon \to 0} \int dx_1 \ldots \int dx_n \ T_\epsilon (x_1 \ldots x_n) \sigma(x_n - x_{n-1}, \lambda) \cdot \psi(x_1 \ldots x_n),$$

where $\mathcal{O}(\lambda)$ = limit $\lambda \to 0$ of the regular part of analytic continuation.

We can write symbolically:

$$\int_{x_{n-1}} dx_n \ T(x_1 \ldots x_n) \phi(x_1 \ldots x_n)$$

$$= \mathcal{O}(\lambda) \lim_{\epsilon \to 0} \int dx_n \ T_\epsilon (x_1 \ldots x_n) \ \sigma_\epsilon (x_n - x_{n-1}, \lambda) \phi(x_1 \ldots x_n)$$

For any pair x_n, x_j of points (i.e. vertices, in configuration space) we introduce a similar distribution

$$\sigma(x_n - x_j, \lambda_{nj}) , \qquad j = 1, 2, \ldots n-1 ,$$

and set, with an obvious extension of the above procedure,

$$\int_{x_1 \ldots x_{n-1}} dx_n \, T(x_1 \ldots x_n) \phi(x_1 \ldots x_n)$$

$$= S\left[\mathcal{O}(\lambda_{n1} \ldots \mathcal{O}(\lambda_{n\ n-1})\right] \frac{1}{n} \lim_{\epsilon \to 0^+} \int dx_n$$

$$\cdot \left[T_\epsilon(x_1 \ldots x_n)\phi(x_1 \ldots x_n) + \sum_{j=1}^{n-1} (x_j \leftrightarrow x_n)\right]$$

$$\cdot \ \sigma_\epsilon(x_n - x_1, \lambda_{n1}) \sigma_\epsilon(x_n - x_{n-1}, \lambda_{n\ n-1}) \ ,$$

where $(x_j \leftrightarrow x_n)$ means $T \epsilon \phi$ with x_j and x_n exchanged, and S is the symmetrizer over the operations $\mathcal{O}$.

Iteration gives, finally, the definition of the multiple F.P. integral:

$$\int_{(x_1 \ldots x_n)} dx_1 \ldots dx_n \, T(x_1 \ldots x_n)\phi(x_1 \ldots x_n) \equiv$$

$$\int dx_1 \int_{x_1} dx_2 \ldots \int_{x_1 \ldots x_{n-1}} dx_n \, T(x_1 \ldots x_n)\phi(x_1 \ldots x_n)$$

$$= \mathcal{O}(\{\lambda\}) \lim_{\epsilon \to 0} \int dx_1 \ldots \int dx_n \, T_\epsilon(x_1 \ldots x_n) \sigma_\epsilon(x_1 - x_2, \lambda_{12})$$

$$\ldots \sigma_\epsilon(x_{n-1} - x_n, \lambda_{n-1,n}) \phi(x_1 \ldots x_n)$$

where now S is the symmetrizer over the variables x in the product $T \epsilon \phi$,

$\{\lambda\}$ $\equiv$ set of $n(n-1)/2$ complex variables λ_{ij} $i \neq j = 1, 2, \ldots, n$ and

$$\mathcal{O}(\{\lambda\}) = \mathcal{O}(\lambda_{12}) S\left[\mathcal{O}(\lambda_{31})\, \mathcal{O}(\lambda_{32})\right] \cdots$$
$$\cdots S\left[\mathcal{O}(\lambda_{n1}) \cdots \mathcal{O}(\lambda_{n\, n-1})\right].$$

Reference (3) contains the proof that this prescription satisfies our axioms. The following notational conventions will be useful in the sequel $\varphi \in S(R^{4n})$:

$$\int_{x_{n-1}} dx_n \, T(x_1 \ldots x_n)\varphi(x_1 \ldots x_n)$$
$$= \lim_{\varepsilon \to 0} \int dx_n T_\varepsilon(x_1 \ldots x_n)\varphi(x_1 \ldots x_n)$$
$$- D_{x_{n-1}}^{x_n} \; T(x_1 \ldots x_n)\varphi(x_1 \ldots x_n) , \qquad (5.9)$$

$$\int_{x_1 \ldots x_{n-1}} dx_n \, T(x_1 \ldots x_n)\varphi(x_1 \ldots x_n)$$
$$= \lim_{\varepsilon \to 0} \int T_\varepsilon(x_1 \ldots x_n)\, \varphi(x_1 \ldots x_n) dx_n$$
$$- \sum_{h=1}^{n} D_{x_h}^{x_n} \; T(x_1 \ldots x_n)\varphi(x_1 \ldots x_n).$$

(5.9) and (5.10) have on the l.h.s. expressions which are well defined and on the r.h.s. expressions whose "meaning" is only formal (unless additional regularizing parameters are introduced); the latter serve to introduce the quantities

$$D_{x_j}^{x_i} \; T(x_1 \ldots x_n)\varphi(x_1 \ldots x_n) ,$$

which will denote in fact the ambiguous or divergent parts that arise, if ordinary integrations are made, at the confluence $x_i \rightarrow x_j$, all other variables being kept different from x_i, x_j. Equation (5.10) is equivalent, in the familiar non-rigorous way of handling these matters, to the definition of the convergent renormalized expression on the l.h.s. by subtraction of divergent parts $D_{x_j}^{x_i}$, from the divergent integral $\int$ on the r.h.s.

In conclusion, the following important property must be explicitly noted:

$$D_{x_j}^{x_i}\; T(x_1 \ldots x_n)\varphi_1(x_1 \ldots x_n) = D_{x_j}^{x_i} T(x_1 \ldots x_n)\varphi_2(x_1 \ldots x_n)$$

if φ_1 and φ_2 asymptotically coincide when $x_i \rightarrow x_j$; this property expresses the quasi-locality of divergent contributions in x-space, and holds for any F.P.I. rule which satisfies our axioms (it is a consequence of axioms a) and b).

6. RENORMALIZATION

6.1 Another Spurion Model

Of this we give, for the sake of brevity, only the shortest mention. Consider a fermion field having scalar interaction with a spurion field, (constant propagator, b, in x-space). Then as is shown in detail in reference 3), the renormalized propagator, after some simplifications, can be expressed completely as

$$K_N = K\begin{pmatrix} x_1 \ldots x_N \\ y_1 \ldots y_N \end{pmatrix}$$

$$= \frac{1}{\sqrt{2\pi b}}\int_{-\infty}^{\infty} dx\; e^{\frac{x^2}{2b}}\; e^{i(1 - \frac{gx}{m})^4 \log(1 - \frac{gx}{m})^2} \cdot \begin{pmatrix} x_1 \ldots x_N \\ y_1 \ldots Y_N \end{pmatrix}_{m-gx}.$$

This is a function of Hadamard class 2, with an essential singularity at the origin of g.

The perturbative expansion of K_N, if "regularized" with a finite volume Ω of integration in x-space and a momentum cut off on the free propagators (x y), can readily be seen to have a finite radius of convergence, independent of N. The analytic behaviour of the exact solution, with respect to g, at the origin is therefore profoundly different from that of the regularized perturbative expansions.

6.2 The $g\varphi^4$ Theory

Again, only results will be quoted: technicalities and proofs may be found in the cited papers. We

expressions of the ambiguous or divergent parts a, b, c and d (functions of m^2 and λ) which emerge when the transition $\int \longrightarrow \int$ is effected: we are actually interested in the transition between two different F.P.I. rules $\int$ and $\int'$, so that a, b, c and d will be finite and describe the renormalization group.

In this theory as we have seen, the propagators have the formal expansions:

$$K_{2n}(x_1 \cdots x_{2n})$$

$$= \sum_{N=0}^{\infty} \frac{(ig)^N}{N!} \int d\xi_1 \cdots \int d\xi_N \left[x_1 \cdots x_{2n} \xi_1 \xi_1 \xi_1 \xi_1 \cdots \xi_N \xi_N \xi_N \xi_N \right],$$

where

$$[x_1 \cdot\cdot x_{2n}\, \xi_1\xi_1\xi_1\xi_1 \,\cdot\cdot\, \xi_N\xi_N\xi_N\xi_N]$$

$$= \sum_{j=2}^{2n} [x_1 x_j][x_2 \cdots x_{j-1} x_{j+1} \cdot\cdot x_{2n}\, \xi_1\xi_1\xi_1\xi_1 \cdot\cdot \xi_N\xi_N\xi_N\xi_N]$$

$$+ \sum_{j=1}^{N} 4[x_1\xi_j][x_2 \cdot\cdot x_{2n}\xi_1\xi_1\xi_1\xi_1 \cdot\cdot \xi_j\xi_j\xi_j \cdot\cdot \xi_N\xi_N\xi_N\xi_N]$$

is a <u>hafnian</u>[15] and

$$[\xi\eta] = \langle T(\varphi(\xi)\varphi(\eta))\rangle$$

$$= iD^{(c)}(\xi - \eta) = \frac{i}{(2\pi)^4}\int \frac{e^{ik.(\xi-\eta)}}{k^2 - m^2 + i\varepsilon}\, dk$$

is the free propagator, or Feynman function. We shall work exclusively with the branching equation

$$K_{2n}(x_1 \ldots x_{2n}) = \sum_{j=2}^{2n} [x_1 x_j] K_{2n-2}(x_2 \cdot\cdot x_{j-1}\, x_{j+1} \ldots x_{2n})$$

$$+ 4ig \int dx\, [x_1 x]\, K_{2n+2}(xxxx_2 \ldots x_{2n}),$$

referred to before as of the type I - those of the type II would be useful only for a deeper study of the renormalization group.

Since

$$(\Box_\xi - m^2)[\xi\eta] = i\delta(\xi - \eta)$$

implies

$$(\Box_{x_1} - m^2)K_{2n}(x_1 \ldots x_{2n})$$

$$= \sum_{j=2}^{2n} i\delta(x_1 - x_j) K_{2n-2}(x_2 \cdot\cdot x_{j-1} x_{j+1} \cdot\cdot x_{2n})$$

$$- 4gK_{2n+2}(x_1 x_1 x_1 x_2 \cdot\cdot x_{2n}).$$

All formulae reported thus far have only a formal meaning: u.v. divergences and ambiguous expressions such as $[\xi\xi]$ are contained in them, so that we must either give them up as hopeless, or try to convert them into meaningful mathematical expressions. We follow the second alternative and prove that, once suitable steps are taken to accomplish this task, the meaningful equations and quantities thus obtained (i.e. the renormalized ones) are formally connected to the expressions above written by means of transformations of parameters

$$(g,\ m^2) \leftrightarrows (\bar{g},\ \bar{m}^2) \qquad \text{(R)}$$

and of functions:

$$\bar{K}_{2n}(x_1..x_{2n}) = B(\bar{g},\ \bar{m}^2)Z^n(\bar{g},\ \bar{m}^2)K_{2n}(x_1..x_{2n}). \qquad \text{(R')}$$

The new propagators $\bar{K}$ are functions of the new parameters $\bar{m}$, $\bar{g}$ which are completely free (if they exist) from infinite or ambiguous parts. $\bar{K}$, $\bar{m}$, $\bar{g}$ are the renormalized quantities (note, in comparing with the standard theory, that we only eliminate ambiguous and infinite parts: additional finite renormalizations will be necessary if it is further required that $\bar{m}$ and $\bar{g}$ be the physical mass and coupling).

A more detailed discussion would show that it is convenient to relate the definition of F.P. integral to the value which we assign to

$$[\xi\xi] = f(\bar{m}^2) \qquad (\neq \lim_{\eta\to\xi}[\xi\eta]!). \qquad (6.2)$$

The regularized expansion ($\int \to \lceil$) is

$$\overline{K}_{2n}(x_1..x_{2n})$$

$$= \sum_{N=0}^{\infty} \frac{(i\overline{g})^N}{N!} \int d\xi_1 \int_{\xi_1} d\xi_2 .. \int_{\xi_1..\xi_{N-1}} .$$

$$.d\xi_N [x_1..x_{2n} \overline{\xi_1\xi_1\xi_1\xi_1} .. \overline{\xi_N\xi_N\xi_N\xi_N}], \quad (6.3)$$

where, clearly:

$$\int_{\xi_1..\xi_{k-1}} d\xi_k = \int d\xi_k - D_{\xi_1}^{\xi_k} - .. - D_{\xi_{k-1}}^{\xi_k} .$$

and the terms of the hafnian in (6.3) are given by

$$[\xi\eta] = \frac{i}{(2\pi)^4} \int \frac{e^{ik.(\xi-\eta)}}{k^2 - \overline{m}^2 + i\varepsilon} dk$$

if $\xi \neq \eta$, or by (6.2), if two points happen to coincide. (This case is different from that which arises when integration over a variable ξ brings it to range over the value of another variable η - this indeed being the cause of u.v. divergences; but no confusion can possibly arise.)

Define now

$$\overline{K}_{2n+2}^{(x)} (\overline{xxxx_2}..x_{2n})$$

$$= \sum_{N=0}^{\infty} \frac{(i\overline{g})^N}{N!} \int_x d\xi_1 \int_{x\xi_1} d\xi_2 .. \int_{x\xi_1..\xi_{N-1}} d\xi_N .$$

$$.[\overline{xxxx}_2 ..x_{2n}\overline{\xi_1\xi_1\xi_1\xi_1} ... \overline{\xi_N\xi_N\xi_N\xi_N}]. \quad (6.4)$$

It follows formally from (6.3) and (6.4)

$$\overline{K}_{2n}(x_1 \ldots x_{2n}) = \sum_{j=2}^{2n} [x_1 x_j]\, \overline{K}_{2n-2}(x_2 \ldots x_{j-1} x_{j+1} \ldots x_{2n})$$

$$+ 4ig \int dx [x_1 x]\, \overline{K}^{(x)}_{2n+2}(\overline{xxx}x_2 \ldots x_{2n}) \tag{6.5}$$

whence, since

$$(\Box_\xi - \overline{m}^2)[\xi \eta] = i\delta(\xi - \eta)$$

it follows

$$(\Box_{x_1} - \overline{m}^2)\, \overline{K}_{2n}(x_1 \ldots x_{2n})$$

$$= i \sum_{j=2}^{2n} \delta(x_1 - x_j) \overline{K}_{2n-2}(x_2 \ldots x_{j-1} x_{j+1} \ldots x_{2n})$$

$$- 4\overline{g}\, \overline{K}^{(x_1)}_{2n+2}(x_1 x_1 x_1 x_2 \ldots x_{2n}) \,. \tag{6.6}$$

Recourse to the formal perturbative expansion is inessential: our task is to deal with equations (6.5) and (6.6); the perturbative series helps to visualize the connections with the non-renormalized theory. Equation (6.6) with the initial condition that $\overline{K}_2(xy) \to [xy]$ when $\overline{g} \to 0$ defines (to within a constant factor, e.g. $\overline{K}_0$) the renormalized propagator, as a consequence of (6.4) which reads: take only the finite parts at all successive integrations, so that any divergences originating because of confluences with x are thereby suppressed (note that here the triplet xxx in the hafnians or in $\overline{K}$ denotes points which are and stay coincident from the beginning! The propagator $\overline{K}$ defined by equation (6.5) (if it exists) has neither ambiguities nor divergences.

A straightforward, though heavy, combinatorial work (which is rendered possible by a full use of the techniques developed at the beginning) permits one to obtain the wanted transformations (R) and (R′) between unrenormalized and renormalized equations or, better, between equations and propagators renormalized with different F.P.I. prescriptions. We shall write, for short, $\int$ and $\int'$ ($\int \equiv \int'$ if we wish so). The result of this work is that one finds

$$K^{(x_1)}_{2n+2}(\overline{x_1 x_1 x_1} x_2 \dots x_{2n}) = (1+c) K_{2n+2}(x_1 x_1 x_1 x_2 \dots x_{2n})$$

$$+ \left[a - 3(1+c)[x_1 x_1]_{\bar{m}^2} + 3F(m^2) \right] \bar{K}_{2n}(x_1 \dots x_{2n})$$

$$- b \Box_{x_1} \bar{K}_{2n}(x_1 \dots x_{2n}) \qquad (6.7)$$

that is, if we call $d = a - 3(1+c)[x_1 x_1]_{\bar{m}^2} + 3F(\bar{m}^2)$ and substitute (6.7) in (6.6)

$$(1 - 4\bar{g}b)\Box_{x_1} - (\bar{m}^2 - 4\bar{g}d) \; \bar{K}_{2n}(x_1 \dots x_{2n})$$

$$= i \sum_{j=2}^{2n} \delta(x_1 - x_j) \, \bar{K}_{2n-2}(x_2 \dots x_{j-1} x_{j+1} \dots x_{2n})$$

$$- 4\bar{g}(1+c) \, \bar{K}_{2n+2}(x_1 x_1 x_1 x_2 \dots x_{2n}). \qquad (6.8)$$

It is at this point trivial to verify that, with the definitions:

$$\overline{K}_{2m}(y_1 \ldots y_{2m}) = B(\overline{g}, \overline{m}^2)\, Z^m(\overline{g}, \overline{m}^2) K_{2m}(y_1 \ldots y_{2m}) \; , \tag{6.9}$$

$$Z(\overline{g}, \overline{m}^2) = \frac{1}{1 - 4\overline{g}b} \; ,$$

$$m^2(\overline{g}, \overline{m}^2) = Z(\overline{m}^2 - 4\overline{g}d) \; ,$$

$$g(\overline{g}, \overline{m}^2) = Z^2(1 + c)\overline{g} \; , \tag{6.10}$$

(6.8) reduces to (6.1), as it was our task to prove.

The quantity B cannot be determined with this procedure; it is instead determined by that used in references (13) and (15). This fact is irrelevant here, since we are mainly interested in the Green functions K_{2n}/K_0 which are free from it.

We can see, thus, that the essence of the problem is the same as was already shown in the example given in section 5.1.

We can go no further here; but it should be evident that our procedure is graph independent and can be applied directly to the branching equations. (It is, in fact, being applied with success to truncated field theories.)

On the one hand the use of F.P.I's permits us to forget about the problem of renormalization entirely, on the other hand it is possible with this technique to describe fully the structure of the renormalization group. This latter work we have already begun.

REFERENCES

1) E.R. Caianiello, Nuovo Cim. 10, 1634 (1953); 11, 492 (1954). Max Planck Festschrift (1958).

2) E.R. Caianiello and A. Buccafurri, Nuovo Cim. 8, 170 (1958).

3) M. Marinaro and F. Guerra, Nuovo Cim. 42, 285 (1966); 60, 756 (1969).

4) E.R. Caianiello, M. Marinaro and F. Guerra, Nuovo Cim. 60, 713 (1969).

5) F. Esposito, U. Esposito and F. Guerra, Nuovo Cim. 60, 772 (1969).

6) T. Carleman, "Les fonctions quasi analytiques", (Paris, 1926).

7) M. Denjoy, Compt. Rend., 173, 1329 (1921).

8) J. Hadamard, "Lectures on Cauchy's Problem", (New York, 1962).

9) M. Gevrey, Compt. Rend., 152, 1564 and Thesis.

10) E.R. Caianiello and S. Okubo, Nuovo Cim. 17 355 (1960).

11) E.R. Caianiello and A. Campolattaro, Nuovo Cim. 20, 422 (1961).

12) F.J. Dyson, Phys. Rev. 75, 486, 1736 (1949).

13) E.R. Caianiello and M. Marinaro, Nuovo Cim. 27, 1185 (1963).

14) N.N. Bogoliubov and D.W. Shirkov, "Introduction to the Theory of Quantized Fields", (New York, 1959).

15) E.R. Caianiello, Nuovo Cim. 13, 637 (1959); 14, 185 (1959).

16) E.R. Caianiello, A. Compolattaro and M. Marinaro, Nuovo Cim. 38, 1777 (1965).

INTRODUCTION TO BRÜCKNER THEORY TO NUCLEAR MATTER

L.J. BOYA
UNIVERSITY OF VALLADOLID
SPAIN

1. INTRODUCTION

The two main problems in nuclear physics are the knowledge of nuclear forces and the determination of the properties of actual nuclei from these forces; both problems are as central today in nuclear physics as they were in 1932, when the nuclear constituents were first established. Of course, the first problem is really the fundamental one, and indeed it can be pushed to a deeper level, namely, by extracting enough information for the nuclear problem from the properties of hadrons (nucleons, π-mesons, vector mesons etc.)

The Brückner nuclear theory, which will be the subject matter of these lectures, intends however to attack the second problem, namely to deduce the properties of actual nuclei (first in an unrealistic simplified case, viz. that of nuclear matter, a concept to be explained later on, then through the study of finite nuclei) from the pertinent two-body (or perhaps three-body) data. In its original form the Brückner theory was essentially an improvement of the original atomic Hartree-Fock method, by the replacement of the two-body potential v by the t-matrix. The Hartree-Fock theory was known, from studies done around 1935-36, not to be a very useful one in nuclear physics, at any rate with the nuclear potentials fashionable at the time. The use of the t-matrix had also the advantage of properly handling singular potentials like those with a hard-core which, since the book of Jastrow (1951), seems to be a part of the nuclear potential, at least in some states. In a broad sense, as Bethe (Bethe 1956) has emphasized, the Brückner theory represents the justification of the nuclear shell model.

The first calculations, done with the Gammel-Thaler potential (Brückner and Gammel 1958) were, as we know today, accidentally successful: they obtained values in sensible agreement with experiment both for the mean density and the specific binding energy for the nuclear matter.

With the 1962 potentials, which reproduce deuteron and two-nucleon scattering data much more accurately, like the Hamada-Johnston or "Breit" potentials, the agreement is no longer so good, and since 1963 Bethe and coworkers have made important improvements on the original Brückner approach, (which we shall also touch on briefly), and the situation today is again fairly satisfactory. Nevertheless we are still far away from having anything like a quantitative picture of fairly complex nuclei.

2. NUCLEAR MATTER

The nuclear matter concept is the first one we want to establish; it is an idealization, not found in nature. Imagine a collection of a large number of nucleons and suppose you can switch off the protonic Coulomb repulsion (more precisely, all the electromagnetic interactions); surface effects will also be negligible because of the number of nucleons being very large. We believe that this conglomerate would have an equilibrium ground state with a density and an average binding energy per particle given by the values

$$-\mathcal{E} = B/A \simeq 15 \text{ MeV} \quad (2.1)$$

$$\rho \simeq 0.22 \text{ nucleons/fermi}^3 \quad (2.2)$$

These values are obtained from the binding energy per nucleon and from the density found in actual nuclei, duly interpreted by means of the Weizäcker semiempirical formula. The fact that the binding energy of nuclei increases linearly with A is known as <u>saturation</u>, and it is an important feature of the nuclear forces.

This imaginary substance we call <u>nuclear matter</u>, and it is approached in the interior of very heavy nuclei. Proper care should be taken, however, in identifying the interior of actual (even heavy) nuclei with nuclear matter: even for $A \simeq 250$ the nucleus is "as much surface as interior".

The first aim of a nuclear theory should be, therefore, to explain these two main features of nuclear matter from the internucleon potential; in particular to show why nuclear matter should be stable, and not for example a collapsed structure. The saturation problem, a very acute one even in the early fifties, is nowadays probably solved through the appropriate combination of exchange forces and the hard-core potentials (and perhaps, as a minor effect, many-body forces).

3. HARTREE FOCK THEORY

As an introduction to Brückner theory we recall briefly the Hartree-Fock theory (1928). Suppose a system of A identical fermions interact through two-body forces; the Hamiltonian is

$$H = \sum_{i=1}^{A} \frac{p_i^2}{2m} + \sum_{i<j} v(i,j) \tag{3.1}$$

The Hartree-Fock approximation is defined as the best replacement of the Hamiltonian (3.1) by a sum of one-body Hamiltonians

$$H_o = \sum_{i=1}^{A} \left\{ \frac{p_i^2}{2m} + U(i) \right\} \equiv \sum_{i=1}^{A} h(i) \tag{3.2}$$

taking statistics properly into account; the Hartree-Fock equations (to be written down immediately) and the Slater determinantal character of the wave function

$$\emptyset_{[\alpha]} = A!^{-\frac{1}{2}} \det u_\alpha(i) \tag{3.3}$$

$$h\, u_\alpha = \varepsilon_\alpha u_\alpha \tag{3.4}$$

are implicit in this definition of the Hartree-Fock theory; here α runs through a complete set of eigensolutions of (3.4) and by $[\alpha]$ we mean a certain pattern of α's. The "goodness" of the Hartree-Fock approximation depends on many factors. In the atomic case it is very well justified since the screened central nuclear potential is a good starting point for an iterative calculation of the self-consistent potential U. The success of the nuclear shell model proves also that the central potential idea is perhaps not too bad even for nuclei.

Only very rarely does one achieve self-consistency in practice; what one does is to start from (3.4) with some approximate central potential and stop after a number of iterations. In the atomic case, the <u>correlation energy</u> i.e., the difference between the exact energy and the Hartree-Fock energy is small, and one usually need not worry about higher order corrections. In the nuclear case however, it is crucial to know how the neglected terms behave, (especially for the infinite nuclear matter case) and this leads us to review, briefly, perturbation theory. This will be done now in some detail, as we believe it will perhaps be the most interesting part of the Brückner theory for the many-body theorist; for completeness we first deduce the Hartree-Fock equations. The word "best" in the previous definition of the Hartree-Fock method is meant in the sense that the first-order perturbation is inoperative; indeed if

$$H_1 = H - H_o = \sum (v-U) \tag{3.5}$$

is the perturbing Hamiltonian, we do not pretend $\langle H_1 \rangle = 0$ but rather that U should be given in terms of v:

$$\langle \alpha | U | \beta \rangle = \sum_{\lambda} \left\{ \langle \alpha \lambda | v | \beta \lambda \rangle - \langle \alpha \lambda | v | \lambda \beta \rangle \right\} \tag{3.6}$$

where λ are <u>occupied levels</u>. That is, λ is in the pattern $[\alpha]$ of the starting state (if we are interested in the ground state, then λ is one of the first 4A levels in (3.3) for the nuclear case). This Fock equation (3.6) could be most easily obtained by Slater's variational calculation; the "exchange" term is a direct consequence of the zero order wavefunction (3.3). So, to first order in v, we have the familar formula

$$E^{(1)} = E_o + \Delta E = \sum_{\lambda} \left\{ \langle \lambda | T | \lambda \rangle + \tfrac{1}{2} \langle \lambda | U | \lambda \rangle \right\} \tag{3.7}$$

where $T = \sum p_i^2 / 2m$ is the kinetic energy, and the characteristic factor $\frac{1}{2}$ comes from the $<$ sign in (3.1) and its absence in (3.6).

Equations (3.6) and (3.4) are the Hartree-Fock self-consistent equations, in the same notation as in the paper of Goldstone (1957).

Let us remark that the Hartree method implies a particular (in fact, the best) choice of the one-body potential U; most of what we are going to say is valid for any central potential U, although we shall speak of self-consistency later on.

4. PERTURBATION THEORY

We now apply the standard Rayleigh-Schrödinger perturbation theory to our many-body problem; in the Brückner "Les Houches Lectures"(Brückner 1958) reasons can be seen why this usual form of perturbation theory is preferable over the Brillouin-Wigner approach.

We repeat the starting equations now in the second-quantization formalism, i.e., in the occupation number representation:

$$H_0 = \sum_{\alpha} \varepsilon_\alpha a^+_\alpha a_\alpha \tag{4.1}$$

is the unperturbed Hamiltonian, with

$$\psi_\alpha(\vec{r}) = \langle \vec{r} \mid \alpha \rangle \tag{4.2}$$

as its one-body solutions; the perturbation is, of course

$$H_1 = \sum_{\alpha,\beta\ldots} \langle \alpha\beta | v | \gamma\delta \rangle a^+_\alpha \; a^+_\beta \; a_\gamma a_\delta - \sum \langle \alpha | U | \beta \rangle a^+_\alpha a_\beta \tag{4.3}$$

and the commutation relations for fermions

$$\{a_\alpha, a^+_\beta\} = \delta_{\alpha\beta} \text{ etc.} \tag{4.4}$$

The formulae of the Rayleigh-Schrödinger theory are

$$E = \langle \phi_0 | H_1 | \psi_0 \rangle = E_0 + \Delta E \tag{4.5}$$

where $\emptyset_0$ is the unperturbed, ψ_0 the exact solution for the (supposedly non-degenerate) ground state; ΔE is the energy shift.

$$|\psi_0\rangle = |\emptyset_0\rangle + \sum_{K=1}^{\infty}\left\{\frac{1-P_0}{E_0-H_0}(H_1-\Delta E)\right\}^K |\emptyset_0\rangle \quad (4.6)$$

and the convention

$$\langle\emptyset_0|\psi_0\rangle = 1 \quad (4.6')$$

Here P_0 projects into the ground state $|\emptyset_0\rangle$

$$1 - P_0 = \sum_{n\neq 0}|n\rangle\langle n| \quad (4.7)$$

For example, the first few orders for the energy are

$$E^{(1)} = E_0 + \langle\emptyset_0|H_1|\emptyset_0\rangle \quad (4.8)$$

$$E^{(2)} = E^{(1)} + \langle\emptyset_0|H_1\frac{1-P_0}{E_0-H_0}H_1|\emptyset_0\rangle$$

$$= E^{(1)} + \sum_{n\neq 0}\frac{|\langle n|H_1|0\rangle|^2}{E_0-E_n} \quad (4.9)$$

$$E^{(3)} = E^{(2)} + \langle\emptyset_0|H_1\frac{1-P_0}{E_0-H_0}H_1\frac{1-P_0}{E_0-H_0}H_1|\emptyset_0\rangle$$

$$- \langle\emptyset_0|H_1|\emptyset_0\rangle\langle\emptyset_0|H_1\frac{1-P_0}{(E_0-H_0)^2}H_1|\emptyset_0\rangle \quad (4.10)$$

where in the last term we do not square $1-P_0$ because it is a projector. Of course, (4.10) can be written more explicitly by using (4.7) once and again as we did for $E^{(2)}$,(4.9). The "normalization" term in $E^{(3)}$, i.e. the second term, becomes more involved in further orders; in general, the N-th term is

$$E^{(N)} = E^{(N-1)} + \langle \emptyset_0 | H_1 \left\{ \frac{1-P_0}{E_0 - H_0} H_1 \right\}^N | \emptyset_0 \rangle$$

$$+ \text{ normalizing terms} \qquad (4.11)$$

For our many-body problem we shall use the graphical representation that Goldstone has borrowed from Feynmann, and we shall also follow Goldstone's conventions (Goldstone 1957, Goldstone 1960, Day 1967). As we are dealing with the ground state as an unperturbed starting state (the vacuum), we must distinguish between states above and below the Fermi level; we shall describe presence of particles above the Fermi levels by up-going lines, and absence of particles below the Fermi level (holes) by down-going lines.

Now we have the two-body operator v and the one-body operator U in our perturbing Hamiltonian (recall (3.5)). Typical matrix elements are, for instance, those shown in Figs. 1 and 2.

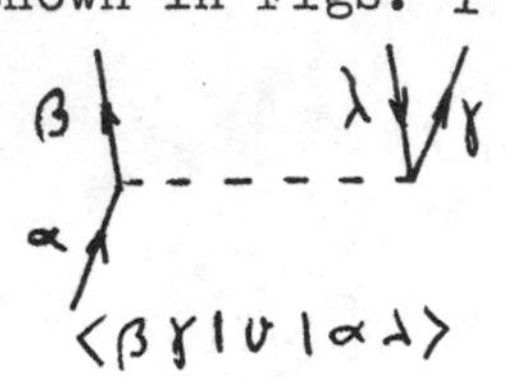

Figure 1

Two-body matrix element with two particles β, λ and two holes α, γ.

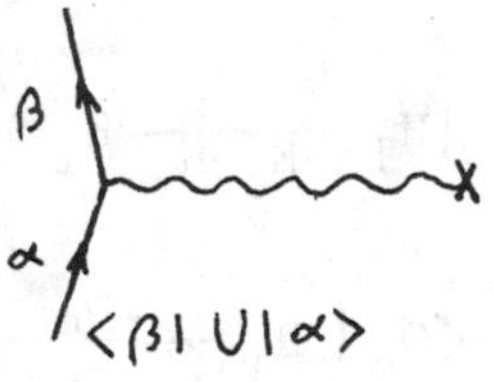

Figure 2

One-body matrix element with two particles; the cross X means the "external" potential U

There is not full symmetry between particles and holes (for one thing there is only a finite possibility for holes), and, as Goldstone emphasizes, there are situations not corresponding to the usual positron theory, for instance that depicted in fig. 3.

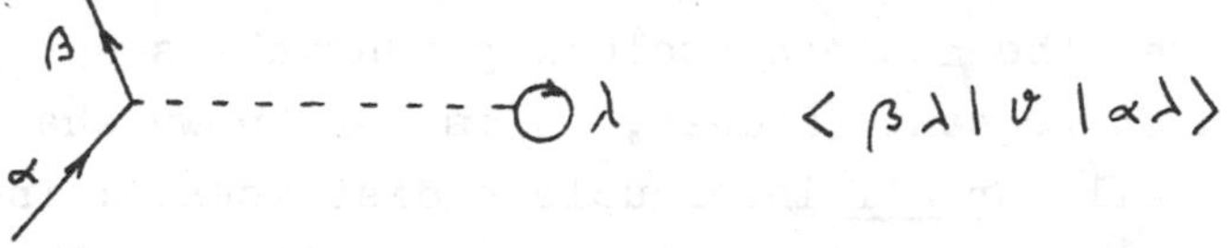

Figure 3. Two-body matrix elements with "passive" unexcited (i.e., below the Fermi level) particles.

This graph represents a particle scattered from the excited state α to the excited state β by the unexcited particle in state λ, which is not changed; these interactions are very important, and should determine the choice of U.

Each possible graph fixes a term in the perturbation expansion; of course, the number of horizontal lines fix the perturbative order, and all graphs should start and end in the vacuum, i.e., in the ground state $\emptyset_0$; for example, those in fig. 4.

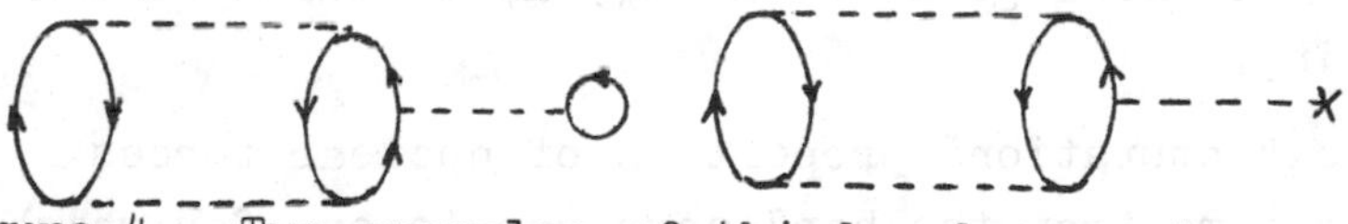

Figure 4. Two examples of third-order graphs are two third-order graphs, the first with a v interaction with passive unexcited particles, the second with a U interaction.

The rules for extracting the contribution from a particular diagram are very simple and we refer to the article of Day (1967); all one has to do is write (and calculate!) the matrix elements of v and U and put the appropriate energy denominators (factors $(E_0 - H_0)^{-1}$ in (4.6)). We leave the matter at the moment and will say more about it in the next lecture; now we want to digress a bit on the internucleon potential v.

5. NUCLEAR POTENTIALS

Of course, the nucleon-nucleon potential is not perfectly known at present (indeed, it is not known whether it exists at all for all internucleon distances!); intelligent guesses, nucleon-nucleon data and meson theory even in its actual rudimentary form have already yielded some results:

1) the "long range" part ($r \gg ƛ_\pi$) of the force is given quite accurately by the one-pion exchange potential (OPEP); this is, of course, the Yukawa tail with the usual spin- and tensor-dependent parts.

2) the "intermediate region"($r \lesssim ƛ_\pi$) is dominated by the one-boson exchange models (OBEM) ($\rho, \omega, \phi, \sigma$?) and perhaps non-resonant two-pion exchange potential TPEP (whose explicit form is still under discussion).

3) the "inner region" ($r \ll ƛ_\pi$) about which we don't know anything.

4) the "saturation" properties of nuclear forces must mainly come from the hard-core repulsion, for which there is some experimental (isotropic p-p scattering) and perhaps theoretical (vector meson exchange) evidence, and for the repulsion in some (mainly odd) states (exchange forces). The hard core has a long history, but recently some "soft core" potentials have again come to the fore. We shall be discussing some of these potentials later, in connection with results from the Brückner-theory calculations; figs. 5 to 8 exemplify some of the more fashionable potentials

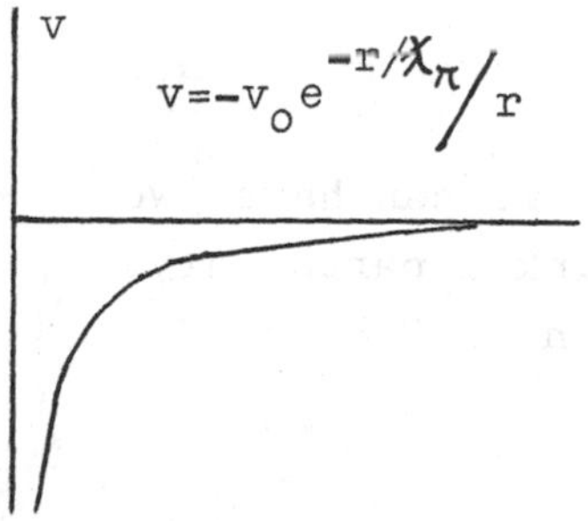

Figure 5. Yukawa potential

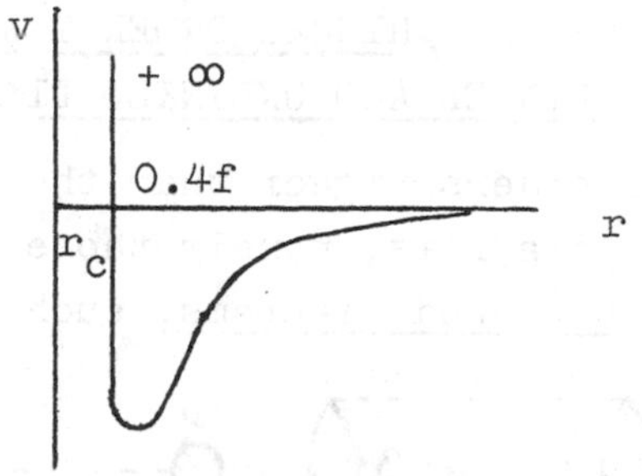

Figure 6. Yukawa + hard core

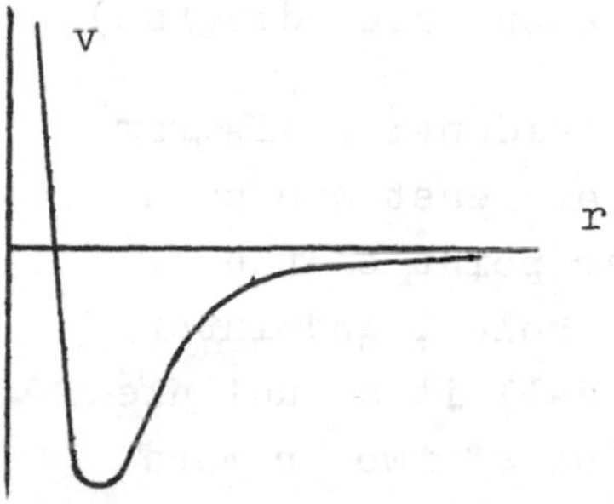

Figure 7. A typical form of "soft-core" potential

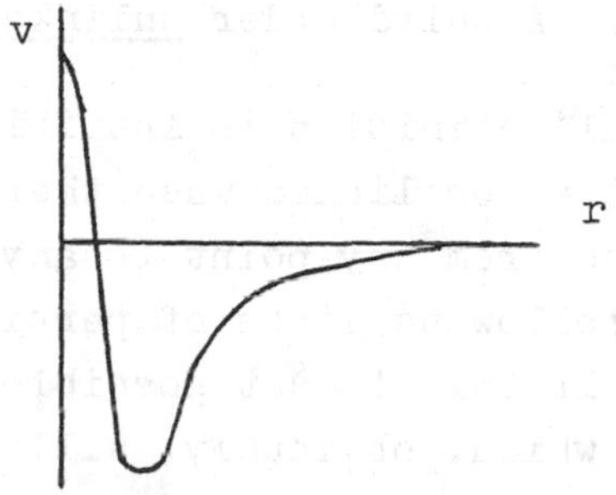

Figure 8. Another typical "soft-core" potential

These figures, of course, mean the potential in some states only; for instance, there are states with only repulsion, and it is not clear if the core (whether hard or soft) is present in all the states; most probably this is not so.

6. HIGHER ORDER TERMS: LINKED AND UNLINKED DIAGRAMS

For orders higher than the second, we can have two types of diagrams, namely those with linked parts only, and the unlinked diagrams, such as shown in Fig. 8

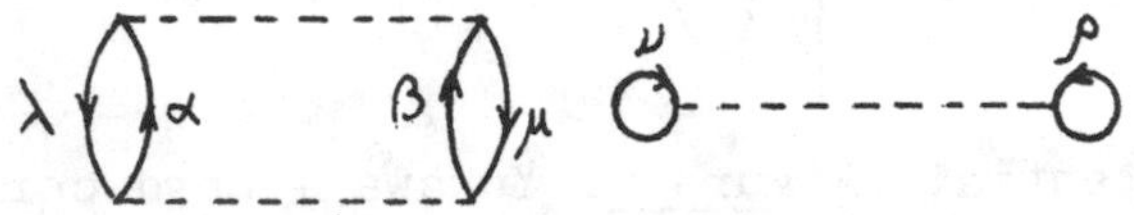

Figure 8. A third order unlinked (disconnected diagram)

The formal definition is intuitively evident: a diagram is connected or linked when there is at least one path which goes from any point to any other point of the diagram following lines of particles, holes, and interactions. If this is not possible, we call it an unlinked diagram, which, obviously, will consist of two or more linked parts. The appearance of these disconnected or unlinked diagrams is very bad, because in the limit of very many particles such a diagram contributes a power higher than the first in the number of particles, therefore the energy per particle does not, in these unlinked diagrams, tend to a constant. The solution to this trouble in the Rayleigh-Schrödinger perturbation theory is contained in the "normalization terms" we mentioned previously. One can show that these terms exactly cancel all the disconnected diagrams, thereby leaving only connected diagrams, which depend linearly on the number of particles (in every perturbative order), thus making the specific binding energy tend to a constant. Brückner explicitly demonstrated this cancellation in third and fourth order, and Goldstone proved the general "linked cluster formula" which we shall explain later.

Let us now see how this comes about in third order, (Brückner 1958). In the nuclear-matter case, translational invariance implies that the single-particle wavefunctions are just plane waves

$$u_\alpha(\vec{r}) = \varphi_{\vec{k}}(\vec{r}) = \langle \vec{k}|\vec{r}\rangle = \Omega^{-\frac{1}{2}} e^{i\vec{k}\cdot\vec{r}} \qquad (6.1)$$

where Ω is the normalization volume. The ground state wavefunction will be of the form

$$\Phi_{[0]}(\vec{r}_j) = \prod_{i=1}^{A} u_{\vec{k}_i}(\vec{r}_j) \qquad (6.2)$$

disregarding, for the time being, normalization and antisymmetrization. Forgetting, for a moment, the fictitious potential, the first-order energy shift is

$$\Delta E^{(1)} = \langle 0|H_1|0\rangle = \sum_{i\neq j} \langle v_{ij}\rangle \qquad (6.3)$$

where v_{ij} represents half the contribution to the energy from a transition in which particles i,j in momentum states k_i, k_j interact and return to the same two momentum states. Since $\langle v_{ij}\rangle$ has two $\Omega^{-\frac{1}{2}}$ factors, and $\sum_{pairs} = (^N 2) \sim N^2$ terms, we have

$$\Delta E^{(1)} \sim N^2 \Omega^{-1} = N\rho \qquad (6.3')$$

i.e., it is proportional to the number of particles.

For the second-order energy shift we have

$$\Delta E^{(2)} = \langle 0|H_1 \frac{1-P_0}{E_0-H_0} H_1|0\rangle = \sum_{ijkl} \langle 0|v_{ij}(1-P_0)(E_0-H_0)^{-1} v_{ij}|0\rangle$$

$$= 2\sum_{ij} \langle 0|v_{ij}\,(1-P_0)(E_0-H_0)^{-1} v_{ij}|0\rangle \sim Nf(\rho) \qquad (6.4)$$

because the excited particles must be brought back to the same states, otherwise the matrix element would be zero. In terms of intermediate states (6.4) is

$$\langle v_{ij}a^{-1}v_{ij}\rangle = \sum_{k_i, k_{i'}} v_{ij,i'i'} \frac{1}{a_{ij,i'j'}} v_{ij,i'i'} \quad (6.5)$$

where $a^{-1}=(1-P_0)(E_0-H_0)^{-1}$

and

$$v_{ij,i'j'} = \langle ij|v|i'j'\rangle \quad ; \quad a^{-1}|i'j'\rangle = a^{-1}{}_{ij,i'i'} \quad (6.6)$$

In fact, the N-dependence expressed in (6.4) is easily proved here too.

In the third order we have, first of all,

$$\Delta E^{(3,1)} = \langle H_1 a^{-1} H_1 a^{-1} H_1\rangle = \sum_{ijklmn} \langle v_{ij} a^{-1} v_{kl} a^{-1} v_{mn}\rangle \quad (6.7)$$

which can be split in two parts. If <u>none</u> of the ijkl coincide, we have a contribution

$$2 \sum_{ij\neq kl} \langle v_{ij} a^{-1} v_{kl} a^{-1} v_{ij}\rangle \equiv \Delta E^{(3,1,unlkd)} \quad (6.8)$$

this is the "unlinked third order term", which corresponds to the graph in fig. 8; since $ij\neq kl$, v_{kl} does not affect ij; hence

$$\Delta E^{(3,1,unlkd)} = 2 \sum_{kl} \langle v_{kl}\rangle \sum_{ij\neq kl} \langle v_{ij}\, a^{-2}\, v_{ij}\rangle \quad (6.9)$$

and, as each linked part contributes an $N\rho$ factor, we would have

$$\Delta E^{(3,1,unlkd)} \sim (N\rho)^2. \quad (6.10)$$

Now the second contribution to the third-order approximation removes this unwanted N-dependence; in fact

$$\Delta E^{(3,2)} = -\langle 0|H_1|0\rangle \langle 0|H_1 \left(\frac{1-P_0}{E_0-H_0}\right)^2 H_1|0\rangle$$

$$= -2 \sum_{kl} \langle v_{kl}\rangle \sum_{ij\neq kl} \langle v_{ij} a^{-2} v_{ij}\rangle + \ldots. \quad (6.11)$$

where the terms that grow more slowly with N have not been given explicitly. Thus (6.9) contain contributions which exactly cancel one another leaving only the correct linear dependence in N for the third order.

7. THE LINKED CLUSTER EXPANSION

We have followed Brückner in this derivation and in the same reference a qualitative discussion is given showing how the same cancellation takes place in fourth order. Then in 1957 Goldstone produced the general proof, using time-dependent perturbation theory. We will state but not prove Goldstone's result, (Goldstone 1957);

$$E = E_0 + \Delta E = E_0 + \langle 0 | H_1 | \sum_L \left(\frac{1-P_0}{E_0 - H_0} \right)^k | 0 \rangle \quad (7.1)$$

where k=1,2,3... and $\sum_L$ means a summation over linked diagrams only, i.e., diagrams with no separate parts. In other words, Goldstone's result amounts to neglecting both the normalization terms in the Rayleigh-Schrödinger expansion and the unlinked diagrams in the main term.

This "linked-cluster expansion", or Brückner-Goldstone expansion, lies in Goldstone's theorem, which was subsequently proved by Hugenholtz (1958) by means of the resolvent operator, by C. Bloch (1958), Hubbard (1958), and others; for more modern references see Brandow (1967).

We are not going to prove Goldstone's theorem, but let us write out the first-order Goldstone diagrams (we are not assuming yet Hartree-Fock self-consistency). The only three first-order diagrams are those in fig. 9

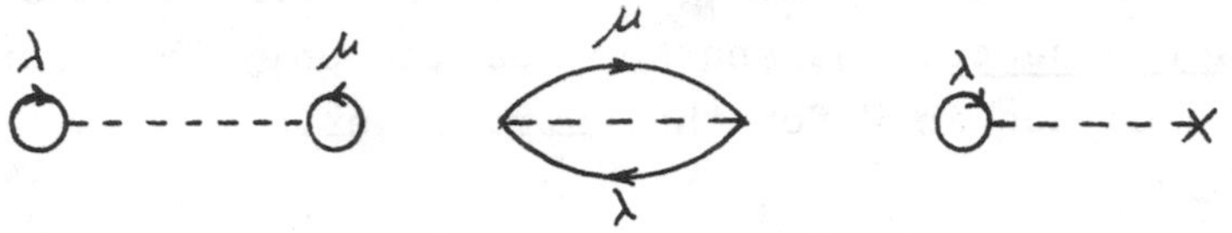

Figure 9. The first-order diagrams: a) the mutual interaction (v-term) of two unexcited passive particles; b) the exchange diagram of a); and c) the Ú-diagram.

The contribution of each diagram to the energy is, respectively

$$\text{a)}\ \tfrac{1}{2}\sum_{\lambda,\mu\le A}\langle\lambda\mu|v|\lambda\mu\rangle$$
$$\text{b)}\ -\tfrac{1}{2}\sum_{\lambda,\mu\le A}\langle\lambda\mu|v|\mu\lambda\rangle \qquad (7.2)$$
$$\text{c)}\ -\sum_{\lambda\le A}\langle\lambda|U|\lambda\rangle$$

The physical reason for the factor $-\frac{1}{2}$ is quite clear - in summing the interactions of pairs of particles, each pair must be counted only once. Of course, to the sum of the first-order diagrams one must add the unperturbed energy E_0,

$$E_0 = \sum E_n = \sum \langle n|T + U|n\rangle \qquad (7.3)$$

which is of course just the sum of the energies of the occupied states. Note now that the U-terms cancel. This cancellation is automatic and is independent of the self-consistency of U. However, the first-order energy

$$E^{(1)}=\sum\langle n|T|n\rangle + \tfrac{1}{2}\cdot\sum\{\langle mn|v|mn\rangle - \langle mn|v|nm\rangle\} \qquad (7.4)$$

still depends implicitly on U, because it is U which determines the zero order states

In this sketchy exposition of the linked-cluster formula we have omitted some minor details; for instance, the treatment of the exclusion principle in excited states, or the precise differences with field-theoretic diagrams; both points are discussed in Goldstone's article or Day's review.

8. THE T-MATRIX EXPANSION

The characteristic feature of the Brückner theory is the use of the t-matrix, instead of the potential, as the basis for a perturbative expansion. The origin of the t-matrix can be seen in the multiple-scattering formalism of Brückner and Watson (1953). There are two reasons for the introduction of the t-matrix: firstly, the t-matrix, computed from the two-body scattering amplitude (with due attention to the exclusion principle), is regular and well-behaved even for singular potentials like the nuclear hard-core, which would make any potential matrix element diverge; secondly, by including at once many "potential" diagrams, it ammounts to a definite improvement on the classical Hartree-Fock method.

In illustrating the t-matrix we shall be following (Day 1967) rather closely . Let us take the connected fourth-order idagram of fig. 10

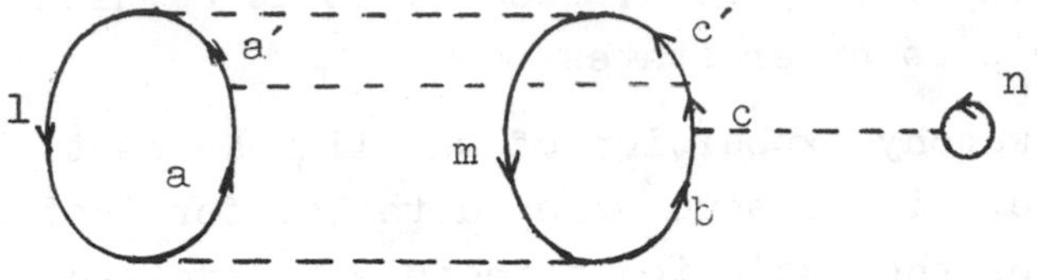

Figure 10. An example of a 4th order diagram for the introduction of t

The matrix element corresponding to the bubble interaction is < n v|bn> . To this diagram we now add an infinite sequence of diagrams, in which the v-interaction is iterated, by introducing ladder interactions between upgoing lines; taking only the (cbn) part of the diagram in fig. 10, this is illustrated in fig. 11.

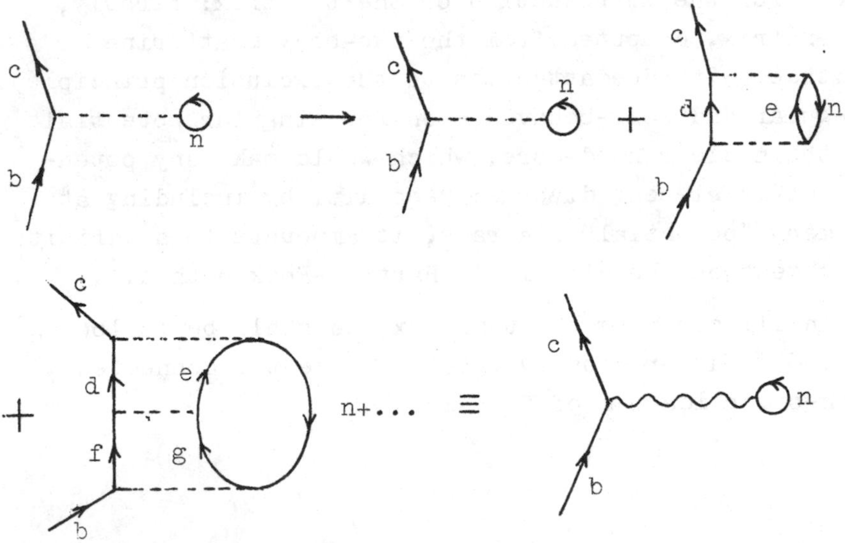

Figure 11. The construction of the t-matrix

The formal definition of t is

$$t = v - v \frac{Q}{e} t. \quad (8.1)$$

t, like v, is a two-body operator; the "exclusion" operator Q just annihilates any two-particle state unless <u>both</u> particles are above the Fermi level (remember the underlined <u>upgoing</u> above); e is the energy of the two-particles minus the initial energy. Expanding t by iteration we get

$$t = v - v \frac{Q}{e} v + v \frac{Q}{e} v \frac{Q}{e} v + \ldots \quad (8.2)$$

which amounts, in fact, to the sum of the diagrams in fig. 11. The integral equation (8.1) amounts to something like the reaction matrix equation in nuclear physics. In a sense, the use of the t-matrix is equivalent to including the exact two body scattering, modified (via the exclusion principle), by the presence of the other particles.

Summing the appropriate sequence of ladder diagrams has led to a single diagram in which one of the v-interaction (dashed line) has been replaced by a t-interaction (wavy line).

The next step, of course, is to replace <u>all</u> the v-interactions by t-interactions. For example, starting from the diagram in fig. 10 we schematize the successive diagrams in fig. 12.

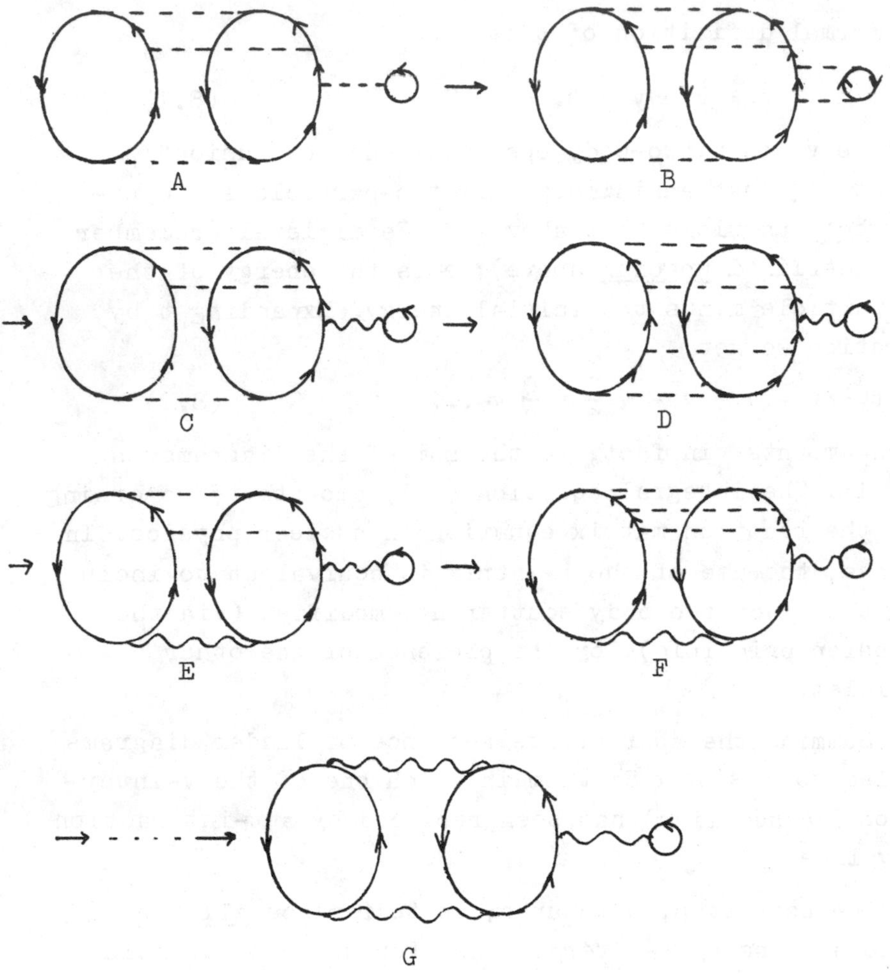

Figure 12. Summation of diagrams starting from the v-form A to the final t-form

The intermediate diagrams have to be drawn with some care; for instance the steps E → F → G should not be done in the sequence shown in fig. 13.

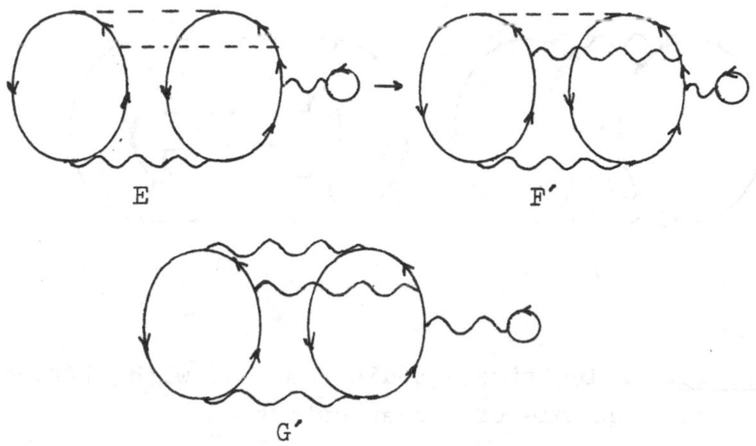

Figure 13. An example of an incorrect summation sequence.

since this would involve double-counting: the diagrams F′ would be counted twice. In fact, we do not need diagrams G′ at all: there should not be adjacent wavy lines between upgoing arrows. To illustrate this point further, we draw in fig. 14 diagram a) which is, by the above criter a redundant one, and give its final form, b), which shows, in fact, that it is a first-order t-diagram. On the other hand, the diagram a) of fig. 15 is not redundant

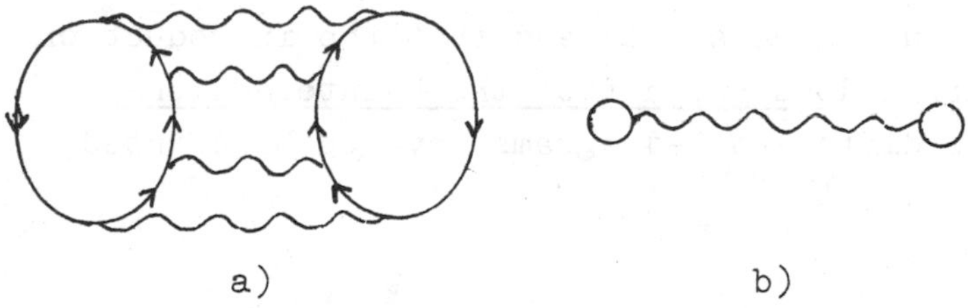

Figure 14. A redundant diagram, a), and its correct form

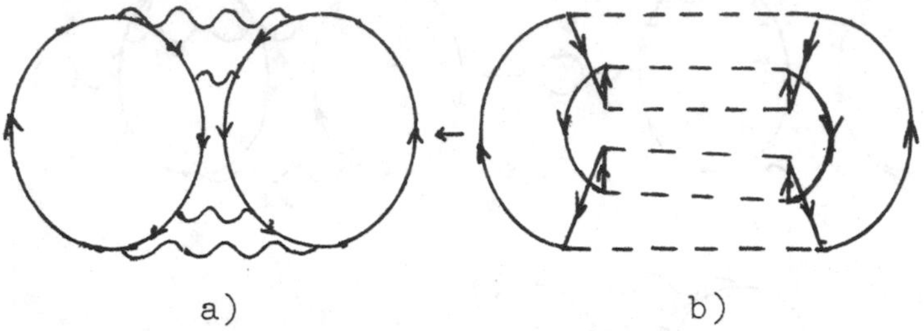

Figure 15. A legitimate t-diagram, a), with different wavy lines, and one of its ancestors, b)

because the wavy lines are between hole lines; fig. 15b) shows one of its ancestors, all of the rest of them being taken into account by adding more "rungs" in the "two-rung" upgoing ladders of this figure.

If we call irreducible a graph in which two v-interactions, between upgoing lines, never appear adjacent to each other, the whole Brückner-Goldstone expansion amounts to

1) Drawing all the connected (linked) diagrams
2) Keeping only the irreducible ones
3) Substituting in them all v-interactions by t-interactions.

In conclusion we should add that the introduction of the t-matrix does not affect the U-interactions at all, and the different U-diagrams have to be counted one by one.

9. IMPLICATION OF MOMENTUM CONSERVATION

Which diagrams does one actually evaluate in a detailed nuclear matter calculation? For nuclear matter, isotropy implies that the zero-order wavefunctions are just plane waves, and that both the one-body kinetic-energy operator $T=p^2/2m$ and the one-body potential U are diagonal in this representation. Of course, the same is true of the internucleon potential, for any decent two-particle interaction! Hence we have momentum conservation at any U or v vertex. This removes many diagrams, as, for example, the one shown in fig. 16

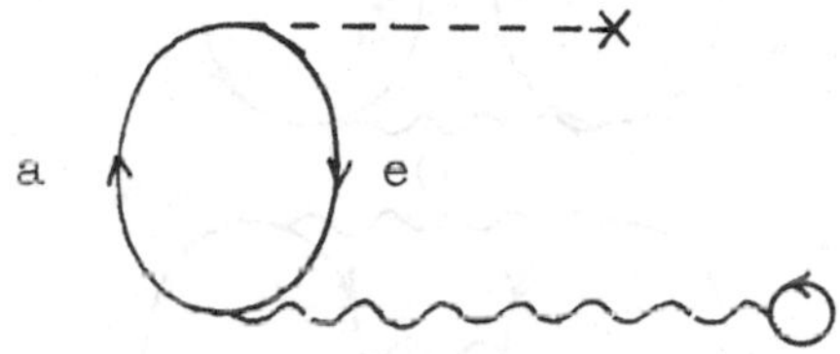

Figure 16. A diagram which does not contribute (in nuclear matter), because momentum is not conserved.

In this diagram the momentum of the intermediate state is $\vec{K}_a - \vec{K}_e$, which cannot be zero because $|\vec{K}_a| > k_F$ and $|\vec{K}_e| < K_F$.

With this proviso it is not difficult to draw the first diagrams which contribute; the first-order (t-)diagrams are, of course, those shown in fig. 17,

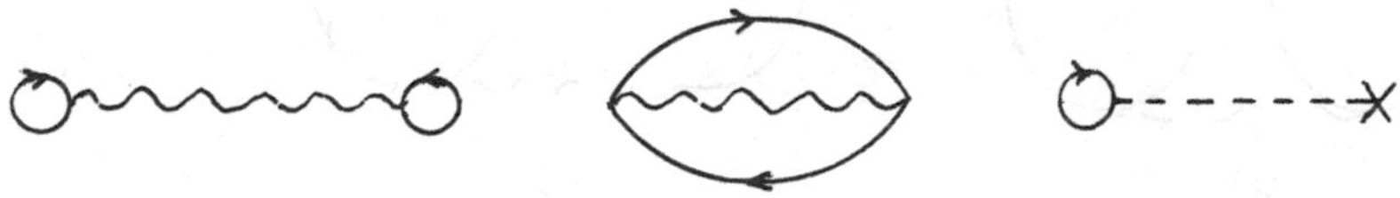

Figure 17. The first-order t-diagrams

that is, the same first-order diagrams of fig. 9, with the wavy lines replacing the dashed v-interactions.

There are no second-order diagrams at all; the reader can easily convince himself that those second-order diagrams which are not redundant in the above mentioned sense, violate momentum conservation.

It is claimed that there are 22 third-order diagrams, some with no U-interactions, as in fig. 18,

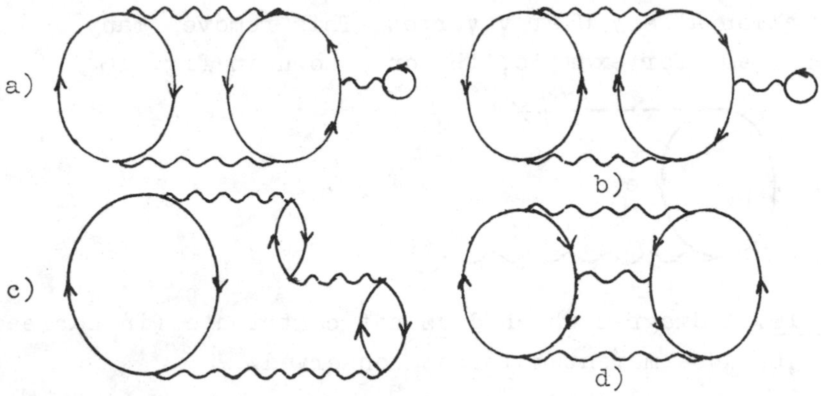

Figure 18. The four types of third-order t-diagrams without U-interactions.

and some with U-interactions included (fig. 19)

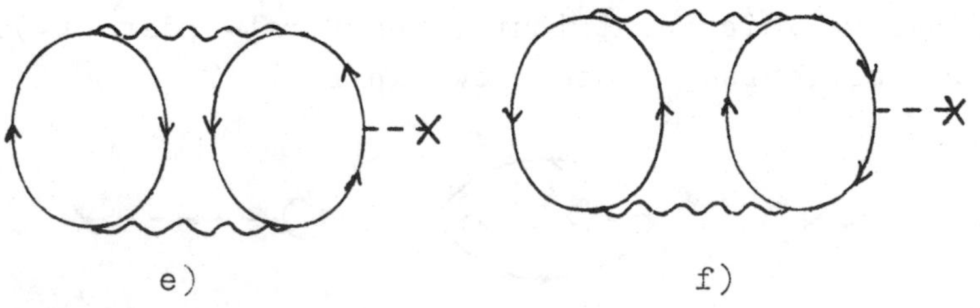

Figure 19. The two types of third-order t-diagrams with U-interactions.

All other third-order diagrams are generated by forming exchange diagrams from those shown here.

So far U still remains at our disposal (remember that the cancellation of the U-terms in first order is automatic for any U), and we may take advantage of this in a (modified) self-consistent form; we shall turn to this point later, but next we shall consider the t-matrix itself and show how it can be obtained - that is, we shall introduce the so-called Bethe-Goldstone equation.

10. THE BETHE-GOLDSTONE EQUATION

Once we have explained what the Brückner-Goldstone expansion amounts to, we must calculate the t-matrix; first, we define a correlated wavefunction by

$$|\psi_{rs}\rangle = (1-\frac{Q}{e}\,t)\,|rs\rangle \qquad (10.1)$$

with the same Q, e and t as before; $|rs\rangle$ is any zero-order two-body state; applying v to (10.1)

$$v|\psi_{rs}\rangle = (v-v\,\frac{Q}{e}\,t)|rs\rangle = t|rs\rangle \qquad (10.2)$$

therefore

$$|\psi_{rs}\rangle = |rs\rangle - \frac{Q}{e}\,v\,|\psi_{rs}\rangle \qquad (10.3)$$

The t-matrix is always computed by first calculating the two-particle correlated state vector $|\psi_{rs}\rangle$; (10.3) is called the Bethe-Goldstone equation (1957). This is an equation for a pair of particles, interacting (to any order) with each other, immersed in a "sea" of all the other particles, which nevertheless interact only with the given pair via the exclusion principle (Pauli operator Q): that is, only momentum components above the Fermi level are available for the scattering.

Let us write (10.3) in a specific representation; the free two-particle wavefunction is

$$\phi_{rs}(\vec{r}_1\vec{r}_2) = \langle \vec{r}_1\vec{r}_2|rs\rangle = \Omega^{-1}\exp(i\vec{k}_r\cdot\vec{r}_1)\exp(i\vec{k}_s\cdot\vec{r}_2)$$

$$= \Omega^{-1}\exp(i\vec{K}_{rs}\cdot\vec{R})\exp(i\vec{k}_{rs}\cdot\vec{r}) \qquad (10.4)$$

where $\vec{R}$ is the center-of-mass coordinate, $\vec{K}_{rs}$ the total momentum, $\vec{k}_{rs}$ the relative momentum and $\vec{r}=\vec{r}_1-\vec{r}_2$. Now for nuclear matter the operators v, Q and e conserve momentum, therefore the Ω, $\vec{K}_{rs}$ and $\vec{R}$-dependence of the correlated state vector $|\psi_{rs}\rangle$ are given by

$$\Psi_{rs}(\vec{r},\vec{R}) = \langle\vec{r},\vec{R}|\Psi_{rs}\rangle = \Omega^{-1}\exp(i\vec{K}_{rs}\cdot\vec{R})\,\Psi_{rs}(\vec{r}) \qquad (10.5)$$

and it is the <u>relative</u> correlated wavefunction $\Psi_{rs}(\vec{r})$ which is important.

Now the t-matrix in the plane-wave representation is

$$\langle pq|t|rs\rangle = \langle pq|v|\Psi_{rs}\rangle = \int\phi^*_{pq}\, v\,\Psi_{rs}\, d^3r_1\, d^3r_2$$

$$= \Omega^{-1}\,\delta(\vec{K}_{pq},\vec{K}_{rs})\,\langle\vec{k}_{pq}|t|\vec{k}_{rs}\rangle \qquad (10.6)$$

where

$$\langle\vec{k}_{pq}|t|\vec{k}_{rs}\rangle = \exp(-i\vec{k}_{pq}\cdot\vec{r})\, v(\vec{r})\,\psi_{rs}(\vec{r})d^3\vec{r} \qquad (10.7)$$

is the t-matrix in k -space; the δ-factor in (10.6) is of course due to overall momentum conservation.

Now it is easy to write the equation for $|\Psi_{rs}\rangle$ in the coordinate representation; this is

$$\Psi_{rs}(\vec{r}) = e^{(i\vec{k}_{rs}\cdot\vec{r})} - \int K(\vec{r},\vec{r}')v(\vec{r}')\Psi_{rs}(\vec{r}')d^3\,r' \qquad (10.8)$$

where the kernel K is given by

$$K(\vec{r},\vec{r}') = \frac{1}{(2\pi)^3}\int d^3k \, \frac{Q(\vec{k},\vec{K}_{rs})}{e(\vec{k},\vec{K}_{rs})} \exp(i\vec{k}\cdot(\vec{r}-\vec{r}')) \tag{10.9}$$

and $Q(\vec{k},\vec{K}_{rs})$ is zero unless

$$|\tfrac{1}{2}\vec{K}_{rs} \pm \vec{k}| > k_F \tag{10.10}$$

for <u>both</u> signs (because $\vec{k}_1,\vec{k}_2 = \frac{1}{2}\vec{K}_{rs} \pm \vec{k}$). If (10.10) holds Q is 1.

Now (10.8) looks very much like the integral equation written down in scattering theory; in fact, making Q=1 (no exclusion principle) and e=kinetic energy operator, the kernel becomes

$$K_{scatt} = \frac{1}{4} \frac{\exp(ik_0|\vec{r}-\vec{r}'|)}{\vec{r}-\vec{r}'} \tag{10.11}$$

where k_0=relative momentum of the initial two-particle state. Substituting (10.11) in (10.8) one gets

$$\psi(r)=\exp(i\vec{k}_0\cdot\vec{r}) - \frac{1}{4\pi}\int\frac{\exp(ik_0|\vec{r}-\vec{r}'|)}{|\vec{r}-\vec{r}'|} v(\vec{r}')\psi(\vec{r}')d^3r' \tag{10.12}$$

which is the exact potential scattering integral equation; of course, the scattering is determined by

$$\psi-\psi_{in} = \psi_{sc} = f(0)\, e^{ikr}/_r \text{(at large r)} \tag{10.13}$$

Now, a characteristic feature of nuclear matter is that ψ_{rs} and ϕ_{rs} are <u>equal</u> at large distances, so there is no scattering at all, i.e., all the "phase shifts" in (10.8) are zero. This phenomena is called the "healing" of the two-body wavefunction; the short-range repulsion makes ψ_{rs} go to zero for small r (in fact, it is exactly zero for $0<r<r_c$ for a hard-core potential). This is a

"wound" in the wavefunction. Now in ordinary scattering, the integral equation (10.12) has a singular kernel (in fact, a prescription must be given to guarantee "outgoing waves" only), but for nuclear matter, there is no singularity in the kernel (10.9) since the energy denominator never vanishes. Therefore, the second term of (10.8) tends to zero at large r, because of the short-range nature of the potential. In other words, no energy conserving momentum components are available for the scattering.

11. THE REFERENCE SPECTRUM METHOD

The difficulty with the Bethe-Goldstone equation lies in the non-local character of the kernel; Brückner originally overcame the difficulty with a partial-wave expansion of the kernel, a numerical evaluation of each wave, and the electronic computation of the whole series; but in 1963, Bethe, Brandow and Petschek developed a much better method, known as the reference-spectrum method. Before going into details, let us first point out the two main advantages.

1) It gives a very simple prescription for a first approximation to t, which turns out to be a fairly accurate one.

2) It also gives a systematic way of improving on this approximation, so enabling one to study quantitatively the higher-order diagrams; this second advantage is a crucial one.

The basic idea of the method is to approximate the odd-looking operator Q/e by a simpler one. If we call $|p\rangle$ the zero-order states, they are eigenstates of momentum, so that

$$e|pq\rangle = \left[E(k_p) + E(k_q)-W\right]|pq\rangle \tag{11.1}$$

$E(k_p)$ is not known unless the potential U(k) is known and this, in turn, involves the self-consistency problem. However, let us approximate the spectral function E(k) by a "reference" spectrum given by

$$E(k)\rightarrow E^R(k) = \frac{k^2}{2m^+} + A_2 \tag{11.2}$$

where A_2 is a constant and m^+ an effective mass. (Of course, (11.2) is equivalent to a very simple momentum-dependent single-particle potential U(k)).

In the coordinate representation the reference-energy operator is given by ($\hbar$=1)

$$e^R = -\frac{1}{2m^+}(\nabla_1^2 + \nabla_2^2) + 2A_2 - W \tag{11.3}$$

The question of how good this approximation is depends naturally on how well a parabolic dependence ($U(k)\propto k^2$) approximates the momentum-dependence of the potential U. The currently accepted choice of U(k) (briefly discussed in section 12) makes $E^R(k)$ approach E(k) fairly well for momenta well above the Fermi level. However, these are the important values, because the operator Q kills momenta less than the Fermi momentum; thus it turns out that a parabolic fit of $E^R(k)$ in the region

$$1.36F^{-1} = k_F < 3F^{-1} \text{ to } 5F^{-1} \tag{11.4}$$

is good enough.

The other approximation in the reference spectrum method is to replace the operator Q by unity. Without going into details, the main reason for this is that the volume of the Fermi sphere is about 1/40 of the volume of the sphere within which important momenta occur (remember $K_F=1.36F^{-1}$). The other reason is that $(e^R)^{-1}$ has a fairly large eigenvalue for the possible cases.

Hence instead of equation (10.2) we now have

$$\psi_{rs} \rightarrow \psi^R_{rs} = \phi_{rs} - \frac{1}{e^R} v \psi^R_{rs} \tag{11.5}$$

Bethe and his coworkers, in their 1963 paper, showed that there is an <u>exact</u> relation between t^R defined obviously by

$$t^R = v - v \frac{1}{e^R} t^R \tag{11.6}$$

and the t-matrix, namely

$$t = t^R + t^R \left[\frac{1}{e^R} - \frac{Q}{e} \right] t \tag{11.7}$$

Thus t^R is a first approximation in a systematic expansion of t. To evaluate the first term, i.e. t^R, we write (11.5) in the form

$$e^R Z^R_{rs} = v \psi^R_{rs} \tag{11.8}$$

where

$$Z^R_{rs} = \phi_{rs} - \psi^R_{rs} \tag{11.9}$$

By separating the volume and center-of-mass dependence of the two-body wavefunctions, we are easily led to

$$(\nabla^2_r - \gamma^2) \zeta^R_{rs} (\vec{r}) = - m^+ v \psi^R_{rs} (\vec{r}) \tag{11.10}$$

where

$$Z^R_{rs} = \Omega^{-1} \exp (i\vec{K}_{rs} \cdot \vec{R}) \zeta^R_{rs} (\vec{r}) \tag{11.11}$$

and

$$\gamma^2 = \tfrac{1}{4} K_{rs}^{\,2} + m^+(2A_2 - W) \qquad (11.12)$$

Equation (11.10) is called the <u>reference wave-equation</u>, and solving it is no more difficult than solving the equation for two-particle scattering. $\psi^R_{rs}(r)$ must vanish inside the (hard) core, and must approach ϕ_{rs} for large r (healing). This fixes the boundary conditions in (11.10), and hence the solution is unique; but we will not discuss the solutions of (11.10) in detail. For a central potential, the usual angular momentum expansion is made, but the method can also be applied to tensor forces.

12. SELF-CONSISTENCY

In nuclear matter, self-consistency in the usual Hartree-Fock sense is trivial as a consequence momentum conservation. Nevertheless, a nontrivial self-consistency requirement arises in the Brückner theory, because the t-matrix depends explicitly on the energy, $t = t(W)$.

The basic idea underlying any definition of U(k) is to cancel as many important t-diagrams as possible; it seems that the most useful definition of U is

$$U(k_m) = \sum_{n \leq A} \left\{ \langle mn \,|\, t(W = E_m + E_n) \,|\, mn \rangle + \text{exchange} \right\} \qquad (12.1)$$

Starting from any U(k), one calculates the state and a new U via t and (12.1); so there is still a self-consistency problem. Also connected with this is the problem of the convergence of the whole Brückner-Goldstone expansion, but here we do not have time to go into these details.

In conclusion, we must say that in its 15 years of existence the Brückner theory has developed to a satisfactory state, and only computational difficulties prevents us from getting direct and precise agreement with the data. Perhaps we should point out the main omission from these lectures, namely the successful treatment that Bethe has given to the three-body clusters (which include more than the third-order diagrams). We finally add that the most recent computations tend to support a soft-core much more than the classical hard-core; whether this will receive support from meson theory, however, remains to be seen.

REFERENCES

1) H.A. Bethe, Phys. Rev. 103, 1353 (1956).
2) J. Goldstone, Proc. Roy. Soc. A239, 267 (1957).
3) K.A. Brückner, in Les Houches Lectures, 1958 (Dunod, Paris, 1959).
4) V.F. Weisskopf, in Les Houches Lectures, 1958, (Dunod, Paris, 1959).
5) J. Goldstone, in Varenna Lectures, 1960.
6) H.A. Bethe, B.H. Brandow and A.G. Petshek, Phys. Rev. 129, 225 (1963)
7) B.P. Day, Rev. Mod. Phys. 39, 719 (1967).
8) R. Rajaraman, H.A. Bethe, Rev. Mod. Phys. 39, 745 (1967).
9) B.H. Brandow, Rev. Mod. Phys. 39, 771 (1967).
10) K.A. Brückner and J.L. Gammel, Phys. Rev. 109, 1023 (1958).

COUPLED BOSON-FERMION SYSTEMS

C.B. DOVER

Institut für theoretische Physik
der Universität Heidelberg

1. INTRODUCTION

A number of interesting physical problems involve the linear coupling of a boson field to a fermion density operator. Such a coupling Hamiltonian H_{int} can be written in the form

$$H_{int} = \lambda \sum_k Q_k \rho_k^+ \qquad (1.1)$$

Here, $\rho_k = \sum_{p\sigma} a^+_{p+k,\sigma} a_{p\sigma}$ is the fermion density operator, $a^+_{p\sigma}$ and $a_{p\sigma}$ being the usual creation and destruction operators for a fermion in momentum state $\vec{p}$ and spin projection σ, respectively. The boson operator is defined by $Q_k = 1/(2\Omega_k)^{\frac{1}{2}}(B_k + B^+_{-k})$ where Ω_k is the free boson energy $(=(\mu^2 + k^2)^{\frac{1}{2}}$ for particles of mass μ), and B^+_k, B_k are creation and destruction operators for mesons of momentum $\vec{k}$. Some of the problems which are characterized by a density coupling of the type (1.1) include

a) the electron-phonon system

b) He^3 - He^4 mixtures

c) the meson-nucleon system, (Chew-Low coupling).

In addition to the boson-fermion coupling (1.1), one may also consider a direct fermion-fermion interaction, for instance the Coulomb force between electrons in example (a).

We shall concentrate most of our attention on the meson-nucleon system. The other cases have been studied extensively from the coupled equations of motion viewpoint, and a number of excellent discussions exist in the literature [1-6].
Although we will not enter into the mathematical details of cases (a) and (b), we will emphasize some of the physical aspects.

2. THE MESON-NUCLEON SYSTEM

For case (c), (1.1) represents the coupling of neutral scalar mesons and nucleons. We will restrict our discussion to this case for simplicity, although the more realistic Chew-Low coupling to pseudoscalar mesons can be treated with equal facility.

Our interest in studying a meson-nucleon coupling of type (1.1) stems from the following considerations.
In theories of nuclear structure, the interaction between two nucleons is usually represented by a static potential, chosen phenomenologically to fit two-nucleon scattering phase shifts. The term "static potential" refers to the fact that the potential exhibits a trivial time dependence $\sim \delta(t)$, i.e., the interaction takes place instantaneously in time.

An example is the well known Yukawa potential

$$V(r,t) = \frac{-\lambda^2}{4\pi} \quad \frac{e^{-\mu r}}{r} \delta(t) \tag{2.1}$$

The Yukawa interaction (2.1) is obtained from the coupling (1.1) by making the static approximation; i.e., the nucleon density operator (a dynamical quantity) is replaced by a prescribed c-number source density. This approximation (infinitely heavy nucleons) is reasonable since $\mu/m \simeq 1/7$ (μ = meson mass; m = nucleon mass). The assumption of a fixed nucleon source density effectively decouples the equations of motion for nucleon and meson degrees of freedom. Except for the spin and isospin degrees of freedom of the nucleon, only the meson variable Q_k is determined by a dynamical equation of motion.

If we do not make the static approximation for the nucleon motion, we are presented with the full complexity of the coupled equations of motion

$$\ddot{Q}_k + \Omega_k^2 Q_k = -\lambda \rho_k$$

$$1/i\,\dot{\rho}^+_{kq\sigma} = \omega^o_{kq}\, \rho^+_{kq\sigma} + \lambda \sum_p Q_p (\rho^+_{k,q+p,\sigma} - \rho^+_{k-p,q+p\,\sigma}) \tag{2.2}$$

where $\rho^+_{kq\sigma} \equiv a^+_{k+q,\sigma}\, a_{k\sigma}$ and $\omega^o_{kq} \equiv \varepsilon^o_{k+q} - \varepsilon^o_k$.

Here $\varepsilon^o_k = k^2/2m$ is the nucleon kinetic energy and we have used the form

$$H_o = \sum_{k\sigma} \varepsilon^o_k\, a^+_{k\sigma}\, a_{k\sigma} + 1/2 \left[\sum_k \left\{P^+_k P_k + \Omega^2_k Q^+_k Q_k\right\}\right]$$

for the non-interacting Hamiltonian, where

$P_k \equiv i(\frac{\Omega_k}{2})^{\frac{1}{2}} (B^+_k - B_{-k})$ is the canonically conjugate variable to Q_k. We treat the meson motion relativistically and the nucleon motion non-relativistically. Indeed, the maximum velocity V_F of a nucleon in a nucleus (at the Fermi surface) is about $V_F/C \sim 0.2$, so the non-relativistic approximation should be adequate.

The basic program is as follows: from the coupled equations (2.2), we wish to extract some information concerning the effective interaction between two nucleons. To achieve this, we must somehow eliminate the meson degrees of freedom from the problem. In Feynman's language, one can always perform, at least in principle, a functional integration over the virtual mesons. In the case of a linear coupling of form (1.1) to the meson field, one would then obtain a fourth order term in fermion operators. This object will be of a non-local character, however, since there will be an integration over time. Physically, this non-locality is just a manifestation of the retarded nature of the fundamental nucleon-nucleon interaction, i.e. the time delay between the emission and absorption of a meson. These time delay effects are completely suppressed when the nucleon-nucleon effective interaction is replaced by a static potential, as for example (2.1).

It will be our goal in the following to explore the influence of retardation effects on problems of nuclear structure. Calculations of nuclear energy spectra and transition rates are usually performed with a static nucleon-nucleon interaction. The question is then: "Why do calculations of nuclear structure based on the shell model with static potentials work so well, when we know that the nucleon-nucleon interaction is mediated by meson exchange and hence is retarded in time?" In the later discussion, we will provide some partial answers to this question.

Let us now illustrate how one eliminates the meson degrees of freedom from the problem, for the case of the linear coupling (1.1). We follow a method developed in references [3,7].
For the identification of the effective interaction, it is most convenient to introduce the one-nucleon Green's function, defined by the time-ordered expression

$$G_{p\sigma}(t-t') = i\langle\Psi_0|T\left\{a_{p\sigma}(t)\, a^+_{p\sigma}(t')\right\}|\Psi_0\rangle \quad (2.3)$$

where $|\Psi_0\rangle$ is the exact ground state of the coupled system. We consider a fermion moving in an infinite Fermi sea (nuclear matter), so that G vanishes unless $a_{p\sigma}$ and $a^+_{p\sigma}$ carry the same momentum and spin indices. The Heisenberg equation of motion for G, corresponding to the neutral scalar coupling (1.1), is found to be

$$(i\frac{\partial}{\partial t} - \mathcal{E}^o_p)\, G_{p\sigma}(t-t')$$

$$= -\delta(t-t') + i\lambda\sum_k \langle T\left\{Q_{-k}(t)a_{p+k,\sigma}(t)a^+_{p\sigma}(t')\right\}\rangle \quad (2.4)$$

The "driving term" on the right-hand side involves the meson degrees of freedom. It satisfies the equation of motion

$$(\frac{\partial^2}{\partial t^2} + \Omega_k^2)\langle T\left\{Q_{-k}(t)\, a_{p+k,\sigma}(t'')\, a^+_{p\sigma}(t')\right\}\rangle$$

$$= -\lambda\langle T\left\{\rho_k(t)\, a_{p+k,\sigma}(t'')\, a^+_{p\sigma}(t')\right\}\rangle . \quad (2.5)$$

The source term on the right-hand side no longer contains meson variables. This is a direct consequence of the fact that the coupling is <u>linear</u> in Q_k. We now can solve (2.5) using the free meson Green's function $D_k(t)$ which satisfies the equation

$$\left(\frac{\partial^2}{\partial t^2} + \Omega_k^2\right) D_k(t-t') = \delta(t-t'). \tag{2.6}$$

We find

$$D_k(t-t') = \frac{1}{2i\Omega_k} e^{i(\Omega_k + i\eta)\,|t-t'|} \; ; \; \eta \to 0^+ \tag{2.7}$$

where we have chosen the boundary conditions $\lim_{t \to \pm\infty} D_k(t) = 0$. Using $D_k(t)$, we may solve (2.5) for $\langle T\{Q_{-k}(t)\, a_{p+k,\sigma}(t'')\, a^+_{p\sigma}(t')\}\rangle$ and substitute the result in (2.4). We then obtain

$$\left(i\frac{\partial}{\partial t} - \varepsilon_p^o\right) G_{p\sigma}(t-t') = -\delta(t-t')$$

$$-i\lambda^2 \sum_k \int_{-\infty}^{+\infty} dt_1\, D_k(t-t_1) \langle T\{\rho_k(t_1) a_{p+k,\sigma}(t)\, a^+_{p\sigma}(t')\}\rangle \tag{2.8}$$

We have now achieved the desired goal of obtaining an equation describing the motion of a single nucleon, but without explicit reference to the meson degrees of freedom. The price we have paid for the elimination of meson variables is the appearance of a non-local operator in time in (2.8). To identify the "effective interaction" of two nucleons, we compare (2.8) with the equation of motion for G obtained by assuming some instantaneous two-nucleon potential V_k in momentum space:

$$\left(i\frac{\partial}{\partial t} - \varepsilon_p^o\right) G_{p\nu}(t-t') = -\delta(t-t')$$

$$+ i \sum_k V_k \langle T\{\rho_k(t)\, a_{p+k,\nu}(t)\, a^+_{p\nu}(t')\}\rangle . \tag{2.9}$$

Comparing (2.8) and (2.9) leads to the identification of the "effective interaction"

$$V_k(t) = - \lambda^2 D_k(t) \tag{2.10}$$

for interactions mediated by neutral scalar meson exchange. An equivalent calculation yields

$$V_k(t) = - (f/\mu)^2 (\vec{\tau}_1 \cdot \vec{\tau}_2)(\vec{\sigma}_1 \cdot \vec{k})(\vec{\sigma}_2 \cdot \vec{k}) D_k(t) \tag{2.11}$$

for the Chew-Low type of coupling to pseudoscalar mesons. The Fourier transform of (2.10) to frequency space is given by

$$V_k(\omega) = \frac{\lambda^2}{\omega^2 - \Omega_k^2} \tag{2.12}$$

We note that $V_k(\omega)$ of (2.12) exhibits an explicit dependence on the frequency ω, (or, equivalently, $V_k(t)$ is explicitly time dependent). As stated before, this reflects the fact that a nucleon may emit a meson which propagates in time, and hence may influence the motion of another (or the same) nucleon at a later time. The fact that $V_k(t)$ exhibits retardation effects is not restricted to our simple model. It will be a general feature of interactions mediated by the exchange of a massive boson, if the boson degrees of freedom are included explicity at the outset.
If we suppress the effects of retardation, i.e., take the limit, $\omega \to 0$, of (2.12), we recover the Fourier transform of the static Yukawa interaction (2.1). Thus, $V_k(\omega)$ of (2.12) may be regarded as the generalization of the usual static one pion exchange potential. However, it should be emphasized that our identification of $V_k(\omega)$ did <u>not</u> depend on the use of perturbation theory. The situation is analogous to the static limit,

where (2.1) represents the exact interaction energy, although it also coincides with the second order perturbation theory result.

Let us now consider some of the properties of $V_k(\omega)$. We first note that $V_k(\omega)$ remains attractive for $\omega < \Omega_k$. The value of ω to be considered depends on the particular problem. For nuclear problems, ω will typically be of the order of the Fermi energy $\varepsilon_F \simeq 40$MeV. Since $\Omega_k > \mu \simeq 140$ MeV, $V_k(\omega)$ will remain predominantly attractive. For the electron-phonon system, the situation is radically different. Here the phonon frequency Ω_k is very small, and $V_k(\omega)$ changes sign at low ω. This fact is important in the theory of superconductivity [8]. Here, in addition to the Coulomb force, one obtains an effective electron-electron interaction similar to (2.12) due to the exchange of lattice phonons.

From (2.12), we can see that for single particle properties in nuclei, the retardation effects will be of order $(\varepsilon_F/\mu)^2 \simeq 0.1$ or smaller. This fact has been verified in detail [7] by calculating the nuclear Hartree-Fock potential using the retarded interaction $V_k(\omega)$ and comparing the result with that obtained with the static Yukawa interaction. To explore the effects of retardation on collective states in nuclei [7], one can solve the Bethe-Salpeter equation for the particle-hole collective eigenfrequencies using $V_k(\omega)$. One finds, not unexpectedly, that the corrections to collective mode energies due to retardation are of order $(\omega_{coll}/\mu)^2$, where ω_{coll} = collective mode frequency. In ^{208}Pb, for example, the 1^- giant dipole state lies at $\simeq 13.5$ MeV, so we have $(\omega_{coll}/\mu)^2 \simeq 0.01$. Thus the effects of retardation will be small.

We can now understand to some extent why a shell model using static potentials provides a rather good description of the low-lying states of nuclear systems. Time delay effects due to the mesonic degrees of freedom will not generally be important due to the smallness of nuclear energies of interest (ε_F and ω_{coll}) compared to the rest mass μ of the exchanged pion. Note that the retardation effects would be of even less importance for exchange of quanta heavier than the π meson. Another way of stating the same result is that the characteristic time for meson motion ($\simeq 1/\mu$) is considerably smaller than a mean time for nucleon motion ($\simeq 1/\varepsilon_F$). Thus to a reasonable approximation, one may neglect effects due to the finite transit time of an exchanged meson.

The interaction $V_k(\omega)$ does not yet contain many-body effects; that is, it does not depend on the Fermi momentum p_F. The identification of $V_k(\omega)$ was based only on the coupled equations of motion, and would be equally valid for the two nucleon system. For a nucleon pair embedded in a many-body system, the Pauli principle excludes scattering states which would be available to an isolated pair. A dependence on the many-body parameter p_F enters through the "screening" of the free space interaction between two nucleons. The "screening" effect has been explored in great detail for the case of the electron gas [1]. Physically, each electron is surrounded by a "polarization cloud" as it propagates through the medium. The free space static Coulomb interaction $4\pi e^2/k^2$ is replaced by a retarded interaction of the form $4\pi e^2/k^2\, \varepsilon(k,\omega)$, where $\varepsilon(k,\omega)$ is the "dielectric constant" of the medium. The retardation arises because of the inertia of the polarization cloud which is "dragged along" as the particle propagates through the medium. A similar effect exists in a nuclear system. In this case we then have two sources of time delay effects: 1) finite transit time of the exchanged meson, 2) inertia of the virtual particle-hole pairs in the polarization cloud.

In the limit $\omega \to 0$, the effect of the screening for the nuclear case is just to decrease the coupling constant and increase the range characterizing the nucleon-nucleon interaction in the many-body system, compared to the corresponding values in free space. [9] The screened interaction

$$V_{eff}(k,\omega) = \frac{\lambda^2}{\omega^2 - \Omega_k^2} \frac{1}{\varepsilon(k,\omega)} \tag{2.13}$$

has also been applied to the calculation of quasi-particle damping in nuclear matter [9]. Here one calculates the imaginary part of the nucleon self energy, using an approximate $\varepsilon(k,\omega)$ calculated in Random Phase Approximation [1,9]. The results have been compared with the imaginary part of the nucleon optical potential in heavy nuclei. Good agreement was obtained for the energy dependence of the absorptive part of the optical potential, although this is to a certain extent fortuitous due to the crudeness of the model, which includes no short range correlations due to the exchange of heavy mesons such as ρ, ω etc.

The above considerations concerning retardation effects constitute only a first tentative step in the program of studying meson exchange between nucleons in a nucleus. The simple model discussed here has several obvious deficiencies. For example, it will not produce nuclear saturation, since a nucleon-nucleon interaction of the form (2.12) or (2.13) does not exhibit the strong short-range repulsion necessary for saturation. One could extend the model to include the exchange of heavier mesons, in order to simulate the repulsive character of the nucleon-nucleon force at small separations. The simple model is valuable, however, in that it provides a simple characterization of the order of magnitude of retardation effects in nuclei.

3. A SOLUBLE MODEL

In the previous section, we have explored in a qualitative way some aspects of retarded interactions in coupled fermion-boson systems. The elimination of the boson degrees of freedom which we discussed works for any linear coupling of the form (1.1), and is not peculiar to the meson-nucleon system. However, we have not yet said anything about how one actually solves the coupled equations of motion (2.2), or the equivalent equations relating meson and nucleon Green's functions. In this section, we discuss a simple model. Within the model the normal modes of the system can be explicitly constructed. In addition, one obtains closed form expressions for the meson Green's function and the density-density correlation function, and the analytic structure of these objects can be fully explored.

Before proceeding, it will be valuable to mention some of the simplifying features of other coupled boson-fermion systems. In the electron-phonon problem, the existence of the small parameter $(m_e/M)^{\frac{1}{2}}$ where m_e = electron mass, M = mass of lattice ion, enables one to obtain expressions for the electron and phonon Green's functions. Migdal [4] has shown that the electron-phonon vertex function Γ can be replaced by 1, the corrections being of order $(m_e/M)^{\frac{1}{2}}$ for normal metals and therefore negligible.

In the problem of He^3 - He^4 mixtures, a phase separation excludes high He^3 densities at low temperature. The system thus consists of a few He^3 impurities (fermions) surrounded by a dense background of He^4 atoms. The existence of a small parameter (the He^3 density) again enables one to obtain equations for the normal mode energies and other properties such as the fermion effective mass [5,6]. The motion of fermions is influenced by the boson background in two ways. Firstly, a fermion becomes clothed by a cloud of virtual excitations, which in this case are phonons and rotons. Secondly, the possibility of exchanging a boson excitation leads to an additional retarded interaction between fermions, as in the meson-nucleon and electron-phonon problems. In the He^3 - He^4, the retardation effects are small because the Fermi velocity is much smaller than the velocity of sound in the background. Thus the transit time of a boson excitation which is exchanged between two fermions is much smaller than the characteristic time for fermion motion.

Now let us return to the meson-nucleon system. A linear coupling of the form (1.1) gives rise to density fluctuations about the equilibrium density distribution of the nucleus (assumed to be infinitely large). In the usual nuclear structure calculations, (employing given static nucleon-nucleon interactions), a well defined approximation (Random Phase Approximation - RPA) exists for linearizing the equation of motion determining the density fluctuations [1,10]. Microscopically, this approach constructs normal modes of the system as linear superpositions of particle-hole excitations. We expect that a similar approach will be valid here under certain conditions.

In the framework of the RPA, one retains only those components of the nucleon density which correspond to particle-hole pair creation or destruction:

$$\rho_k^+ \simeq \Big(\sum_{\substack{|\vec{p}| < p_F \\ |\vec{p}+\vec{k}| > p_F \\ \sigma}} + \sum_{\substack{|\vec{p}| > p_F \\ |\vec{p}+\vec{k}| < p_F \\ \sigma}} \Big) a^+_{p+k,\sigma} a_{p\sigma} \qquad (3.1)$$

Terms involving particle-particle or hole-hole scattering are suppressed. Moreover, the density operators $\rho_{kq\sigma} \equiv a^+_{k+q,\sigma} a_{k\sigma}$ are assumed to satisfy boson commutation relations

$$\left[\rho_{pk\sigma}, \rho^+_{p'k'\sigma'}\right] \simeq \delta_{pp'}\, \delta_{kk'}\, \delta_{\sigma\sigma'} \qquad (3.2)$$

Physically, (3.2) implies that a particle-hole pair propagates as a unit, and does not exchange particles or holes with other pairs. This assumption clearly violates the Pauli principle to some extent, but is reasonable if the number of virtual particle-hole pairs is not too large. This is consistent with the assumption that the density fluctuations are small, and do not greatly change the character of the non-interacting Fermi sea; i.e., the coupling is not too strong.

Using (3.2), the coupled equations (2.2) assume the greatly simplified form

$$\begin{aligned} \ddot{Q}_k^+ + \Omega_k^2 Q_k^+ &= -\lambda \sum_p{}' A^+_{pk} \\ \ddot{A}^+_{pk} + {\omega^o_{pk}}^2 A^+_{pk} &= -2\lambda Q_k^+ \omega^o_{pk} \end{aligned} \qquad (3.3)$$

where $A^+_{pk} = \rho^+_{pk} + \rho_{-p,-k}$. The use of the boson approximation has produced a pair of coupled oscillator equations, which are soluble by a straightforward normal mode analysis. The equation determining the eigenfrequencies of the system is obtained by assuming an $\exp(-i\omega t)$ time dependence for Q^+_k and A^+_{pk}. This yields

$$\omega^2 - \Omega_k^2 = 2\lambda^2 \sum_p{}' \frac{\omega^o_{pk}}{\omega^2 - \omega^{o2}_{pk}} \qquad (3.4)$$

where the prime on Σ denotes $|\vec{p}| < p_F$, $|\vec{p} + \vec{k}| > p_F$.

The solutions of (3.4) are of two types. Firstly, for an infinite system, the unperturbed particle-hole energies are still solutions of the coupled system. Physically, this means that particle-hole pairs have no self-energy in this simple model. In addition, a renormalized or "dressed" meson mode appears, which reduces to the bare meson solution Ω_k in the weak coupling limit. There is no collective mode with $\omega \to 0$ as $k \to 0$. This is a consequence of the fact that our model includes only the direct matrix element of a neutral scalar interaction, which is here attractive. An additional solution, if it exists, would be pushed down from its unperturbed energy, and hence strongly damped ("Landau damping") into the particle-hole continuum.

It should be noted that (3.4) has the same form as the dispersion relation for collective oscillations in an electron gas [1], with the Coulomb interaction replaced by the retarded interaction $V_k(\omega)$ of (2.12). Thus this model yields the same $V_k(\omega)$ as the method of elimination of meson variables presented in the first section.

The explicit expressions for the operators ρ^+_{pk} and C^+_k which create particle-hole or dressed meson normal mode excitations are easily constructed as linear superpositions of unperturbed particle-hole excitations[11]. These objects will form a complete set for the expansion of other quantities of interest. For example, the density operator ρ^+_k has an expansion of the form

$$\rho^+_k |\psi_o\rangle = \sum_p C_{pk} \; \rho^+_{pk} |\psi_o\rangle + C_\pi(k) \; C^+_k |\psi_o\rangle \qquad (3.5)$$

where C_{pk} and $C_\pi(k)$ are certain functions which may be easily determined. The weights $|C_{pk}|^2$ and $|C_\pi(k)|^2$ represent the spectral composition of the density fluctuations.

Having constructed the normal modes, one can easily evaluate the exact ground state energy of the system. We first use the commutator

$$[H, B_k] = -\Omega_k \; B_k - \frac{\lambda}{(2\Omega_k)^{\frac{1}{2}}} \; \rho_k \qquad (3.6)$$

(which follows from (1.1)) and its Hermitian conjugate to yield the identity

$$Q_k \; |\Psi_n\rangle = \frac{\lambda}{(H-E_n)^2 - \Omega_k^2} \; \rho_k \; |\psi_n\rangle \qquad (3.7)$$

where $H|\psi_n\rangle = E_n|\psi_n\rangle$. We now take the ground state expectation value of the interaction (1.1), using (3.7) to eliminate the meson coordinates and then (3.5) to express the result in terms of C_{pk} and $C_\pi(k)$:

$$\langle\Psi_0|H'|\Psi_0\rangle = \lambda^2 \sum_{kn} \frac{1}{\omega_\pi^2 - \Omega_k^2} |\langle\Psi_n|\rho_k^+|\Psi_0\rangle|^2$$

$$= \lambda^2 \sum_k \left\{ \frac{|C_\pi(k)|^2}{\omega_\pi^2(k) - \Omega_k^2} + \sum_p{}' \frac{|C_{pk}|^2}{\omega_{pk}^{0\,2} - \Omega_k^2} \right\} \tag{3.8}$$

The energy is seen to be a weighted sum of terms of the form $\lambda^2/\omega^2-\Omega_k^2$, the weights $|C_\pi(k)|^2$ and $|C_{pk}|^2$ representing the normal mode decomposition (3.5) of ρ_k^+.

Within the RPA, one can also obtain the meson and nucleon density Green's functions in closed form. For instance, defining the meson propagator

$$R(k,t) \equiv i \langle\Psi_0| T\{B_k(t)\, B_k^+(o)\} |\Psi_0\rangle \tag{3.9}$$

we can again insert a complete set of states in (3.9) and use (3.5) and (3.6). The result is

$$R(k,t) = \frac{i\lambda^2}{2\Omega_k}\left[\sum_p{}' \left(\frac{1}{\omega_{pk}^0 - \Omega_k}\right)^2 |C_{pk}|^2 \, e^{-i\omega_{pk}^0 t} \right.$$

$$\left. + \left(\frac{1}{\omega_\pi(k) - \Omega_k}\right)^2 |C_\pi(k)|^2 \, e^{-i\omega_\pi(k)t}\right]; \; t>o \tag{3.10}$$

$t<o$; $R(k,t) = 0$. Here $\omega_\pi(k)$ is the dressed meson frequency. Taking the Fourier transform of (3.10), we can calculate the "optical potential" seen by a meson incident on a nucleus within the framework of the model. However, the absorption of mesons by nuclei is known to proceed primarily via two nucleon absorption (πNN), and to depend strongly on the presence of short range correlations in the two nucleon force. Thus our simple model can certainly not explain the experimentally observed absorption of mesons, since it includes only

long range correlations due to the exchange of neutral scalar or pseudoscalar mesons. However, it may be reasonable to use (3.10) to calculate the real part of the meson optical potential for heavy nuclei.

The nucleon density-density correlation function $G_\rho(k,t)$ is also calculable in our model. Defining $G_\rho(k,t)$ by

$$G_\rho(k,t) = \langle\psi_o|\, T\{\rho_k(t)\,\rho_k(0)\}\,|\psi_o\rangle \tag{3.11}$$

we find directly from (3.5) the result

$$G_\rho(k,t) = \sum_p{}' |C_{pk}|^2\, e^{-i\omega^o_{pk}|t|} + |C_\pi(k)|^2 e^{-i\omega_\pi(k)|t|} \tag{3.12}$$

The usual [1)] correlation function S(k,t) is just the $t > o$ part of G_ρ. S(k,t) contains information concerning the dynamical correlation in time of two particle-hole pairs. The so-called "dynamic form factor" $S(k,\omega)$ is given by

$$S(k,\omega) = \sum_p{}' |C_{pk}|^2\, \delta(\omega-\omega^o_{pk}) + |C_\pi(k)|^2\delta(\omega-\omega_\pi(k)) \tag{3.13}$$

for our model. In principle, the excitation spectrum $S(k,\omega)$ of the density fluctuations is measurable in a scattering experiment involving some weakly coupled probe.

It should be noted that both the meson Green's function $D(k,\omega)$ and the density correlation function $G_\rho(k,\omega)$ exhibit poles at the normal mode frequencies ω^o_{pk} and $\omega_\pi(k)$. This is a manifestation of the coupling of meson and nucleon degrees of freedom. For $D(k,\omega)$, the residues at the particle-hole singularities ω^o_{pk}

are proportional to λ^2 for weak coupling, and of course vanish in the non-interacting limit. Similarly, the residue of $G_\rho(k, \omega)$ at the dressed meson pole is $\sim\lambda^2$ for $\lambda^2 \ll 1$. The fact that the boson propagator will contain singularities due to the fermions in a coupled system and vice versa has also been noted in the He^3 - He^4 and electron-phonon problems [6,12].

The simple RPA approximation for treating the coupled meson-nucleus system does not provide a quantitative description of real nuclear systems. However, it is useful as a vehicle by means of which one can obtain the meson and nucleon density Green's functions in closed form, as well as the exact interaction energy and other properties. The emphasis here is on probing the analytic structure of the theory. As such, the model provides a useful illustration of many-body Green's function techniques.

REFERENCES

1) P. Nozieres and D. Pines, "Quantum Theory of Liquids", (Benjamin, Inc. New York, 1966).

2) G. Schultz, "Quantum Field Theory and the Many-Body Problem", (Gordon and Breach, New York, 1963).

3) L. Kadanoff, "The Electron-Phonon Interaction in Metals" in "Lectures on the Many-Body Problem", (Vol. 2), (Academic Press, New York, 1964).

4) A. Migdal, Sov. Phys. JETP 7, 996 (1958).

5) S.V. Tyablikov, Sov. Phys. Solid State 3,2500 (1962).

6) S.T. Stenholm, Physics 38, 608 (1968)

7) C.B. Dover and R.H. Lemmer, Phys. Rev. 165, 1105(1968).

8) H. Frölich, Phys. Rev. 79, 845 (1950).

9) C.B. Dover and R.H. Lemmer, to appear in Phys. Rev.

10) A.M. Lane, "Nuclear Theory", (Benjamin, Inc. New York, 1964).

11) C.B. Dover, Ann.Phys. (N.Y.) 50, 449 (1968)

12) A.M. Badalyan and L.A. Maksimov, Sov.Phys. JETP 23, 518 (1966).

PLASMON RESONANCES IN THE QUANTUM STRONG FIELD LIMIT

NORMAN J. HORING

Department of Physics
Stevens Institute of Technology
Hoboken, New Jersey, U.S.A.

ABSTRACT

The plasmon spectrum is investigated in the quantum strong field limit, ($\hbar\omega_c > \zeta \sim E_F$). We find another set of plasmon resonances in addition to those already discussed in the literature. This new set of resonances is only very mildly damped under appropriate conditions of low wavenumber. Under such low wavenumber conditions, the new resonances are of comparable importance (in terms of excitation amplitude) with other resonances previously reported. The result is that there are two plasmon resonances which lie near each multiple of the cyclotron frequency, except for propagation nearly perpendicular to the magnetic field.

1. INTRODUCTION

The quantum theory of electron gas plasmon propagation in a magnetic field has been considered by several authors [1,2,3] with the use of the random phase approximation (R.P.A.). Our object here is to complete the study of plasmon resonances in a magnetic field which was undertaken in our earlier work [4], and in so doing, to call attention to the existence of a set of resonances which has not yet been discussed in the literature. Specifically, we are interested in carrying out a detailed analysis of plasmon resonances (as well as their damping and excitation amplitudes) in the quantum strong field limit, when the energy separation of the Landau levels ($\hbar\omega_c$) is larger than the chemical potential (Fermi energy) ($\zeta \sim E_{Fermi}$) as well as the thermal energy ($1/\beta = KT$). One set of plasmon resonances in the quantum strong field limit has already been discussed [2], and it associates one resonance with each multiple of the cyclotron frequency $n\omega_c$, which lies close to $n\omega_c$ under appropriate conditions of low wavenumber. We demonstrate the existence of a second set of plasmon resonances, so that there are two resonances which lie close to each $n\omega_c$. The first set of resonances is undamped, and the second set of resonances is only very mildly damped under appropriate conditions of low wavenumber. Under such low wavenumber conditions, the two sets of resonances are of comparable importance in terms of excitation amplitude. (However, in the special case of plasmon propagation nearly perpendicular to the magnetic field, the second set of resonances becomes relatively unimportant, and then there is as usual just one resonance near each $n\omega_c$).

In section 2 a formal derivation of the dispersion relation and damping constant necessary for the examination of the plasmon resonances in the quantum strong field limit is given, and in section 3 the examination of the resonances, as well as their damping and excitation amplitudes, is carried out. The conclusions are summarized in section 4.

2. THE DISPERSION RELATION AND DAMPING CONSTANT

The plasmon dispersion relation in the presence of an arbitrary strong magnetic field has been derived in the random phase approximation in reference 4d*, (equations (3.26) and (3.27). There, we showed that it could be written in the form,

$$1=\frac{4\pi e^2}{p^2}\int_0^\infty d\omega\,\frac{f_0(\omega)}{\hbar^3}\int_{c-i\infty}^{c+i\infty}\frac{ds}{2\pi i}\,e^{s\omega}\,\frac{\pi^{3/2}}{(2\pi)^3}\left(\frac{2m}{s}\right)^{1/2}\frac{\hbar m\omega_c}{\tanh(\hbar\omega_c s/2)}\aleph \tag{2.1}$$

where

$$\aleph=\frac{2i}{\hbar}\int_0^\infty dT e^{-i\Omega T}\exp\left[\frac{-\hbar^2 p_z^2 s}{8m}\right]\exp\left[\frac{-\hbar\bar{p}^2}{2m\omega_c}\coth(h\omega_c s/2)\right] \tag{2.2}$$

$$\cdot\left\{\exp\left[\frac{-p_z^2}{8ms}(2T-i\hbar s)^2\right]\exp\left[\frac{\hbar\bar{p}^2}{2m\omega_c}\,\frac{\cos([\omega_c/2][2T-i\hbar s])}{\sinh(\hbar\omega_c s/2)}\right]-(i\to -i)\right\}$$

The right hand side of this dispersion relation has simple poles at all multiples of the cyclotron frequency in the special case of propagation perpendicular to the magnetic field. A representation of the dispersion relation for arbitrary propagation direction which explicitly corresponds to this may be obtained by

---

*The notation of reference 4d will be maintained here:
$f_0(\omega)$= Fermi-Dirac distribution function,
$\vec{H}$ = magnetic field along z-axis; $\vec{p}=(p_z, \bar{p})$=wave vector;
ζ = chemical potential = Fermi energy E_F; $\beta=(KT)^{-1}$;
ω_p = classical plasma frequency $(4\pi e^2 p/m)^{\frac{1}{2}}$;
ω_c = cyclotron frequency eH/mc; p_F=Fermi wavenumber $(2m\zeta)^{\frac{1}{2}}/\hbar$.

expanding the factor $\exp[y\cos\varphi] = \sum_{n=-\infty}^{\infty} e^{in\varphi} I_n(y)$ in terms of modified Bessel functions I_n (B.H.T.F.[5] II, pg. 7, No. 27): The ensuing T-integral is readily recognized as a complementary error function (Erfc), and thus one has the result,

$$\mathcal{N} = \frac{i\pi^{\frac{1}{2}}}{\hbar}\left(\frac{2ms}{p_z^2}\right)^{\frac{1}{2}} \exp\left[\frac{-\hbar^2 p_z^2 s}{8m}\right] \exp \frac{-\hbar\bar{p}^2}{2m\omega_c} \coth(\hbar\omega_c s/2)$$

$$\sum_{n=0}^{\infty}{}' I_n\left[(\hbar\bar{p}^2/2m\omega_c \sinh(\hbar\omega_c s/2))\right]$$

$$\cdot\Big\{ e^{\hbar\Omega s/2} \exp\left[-m(\Omega - n\omega_c)^2 s/2p_z^2\right]$$

$$\mathrm{Erfc}\left[i(m(\Omega - n\omega_c)/p_z^2 - \hbar/2)(p_z^2 s/2m)^{\frac{1}{2}}\right] - (\hbar \to -\hbar)$$

$$+ e^{\hbar\Omega s/2} \exp\left[-m(\Omega + n\omega_c)^2 s/2p_z^2\right]$$

$$\mathrm{Erfc}\left[i(m(\Omega + n\omega_c)/p_z^2 - \hbar/2)(p_z^2 s/2m)^{\frac{1}{2}}\right] - (\hbar \to -\hbar)\Big\}. \quad (2.3)$$

The prime on $\sum_{n=0}^{\infty}{}'$ indicates that the n=0 term is double-counted, and should be divided by 2. For very strong magnetic fields ($\hbar\omega_c \gtrsim \zeta$; $\hbar\omega_c\beta \gg 1$), it is appropriate to expand* the part of $\mathcal{N}$ involving hyperbolic functions

---

* Such an expansion may be obtained by using the identity, (Watson, J. London Math. Soc. 8, 189 (1938)),

$$\exp\left[\frac{-\hbar\bar{p}^2}{2m_c} \coth(\hbar\omega_c s/2)\right] I_n\left[(\hbar\bar{p}^2/2m\omega_c \sinh(h\omega_c s/2))\right]$$
$$= 2\left(\frac{\hbar\bar{p}^2}{2m\omega_c}\right)^n .$$

in a series of powers of $\exp[-\hbar\omega_c s/2]$. The resulting expression for $\aleph$ may be substituted into the dispersion relation (2.1), and one then finds that the s-integral of (2.1) is given by the inverse Laplace transform of a complementary error function, and this may be evaluated in terms of elementary functions (B.I.T.[5] I, pg. 266, No. 12). Thus, with a bit of manipulation, one obtains the dispersion relation in the form **

$$\exp\left[-\hbar\bar{p}^2/2m\omega_c\right] \sinh\,(\hbar\omega_c s/2)$$

$$\sum_{r=0}^{\infty} \frac{r!}{(n+r)!}\; L_r^n(\hbar\bar{p}^2/2m\omega_c)^{\;2} \exp\left[-(n+2r+1)\hbar\omega_c s/2\right],$$

(L_r^n denotes the Laguerre polynomial).

** In this equation there are two independent summations over the two possibilities of signature $\pm$: On the one hand, $\sum_\pm$ arises from the two possibilities of signature associated with spin; and on the other hand, $\sum_{\pm}{}'$ arises in connection with folding the summation $\sum_{n=-\infty}^{\infty}$ into $\sum_{n=0}^{\infty}$, and it provides an accounting of the negative n terms along with the positive ones-- in so doing the n = 0 term is double-counted and must be divided by 2 (hence the sume $\sum_{n=0}^{\infty} \rightarrow \sum_{n=0}^{\infty}{}'$ carries a prime to indicate this).

$$1 = -\frac{4\pi e^2}{p^2}\frac{m^2\omega_c}{4\pi^2 p_z}$$

$$\sum_{n=0}^{\infty}{}' \sum_{r=0}^{\infty} \left(\frac{\hbar \bar{p}^2}{2m\omega_c}\right)^n \exp\left(\frac{-\hbar \bar{p}^2}{2m\omega_c}\right) \frac{r!}{(n+r)!}\left[L_r^n\left(\frac{\hbar \bar{p}^2}{2m\omega_c}\right)\right]^2 \tag{2.4}$$

$$\sum_{\pm\pm'} \frac{1}{2\hbar^3}\int d\omega' f_0(\omega' + a_{rn}(\pm\pm'))\Big\{\left[\omega' - (m\omega'/2p_z^2)^{\frac{1}{2}}\Theta_n^{\pm'}\right]^{-1}$$

$$-\left[\omega' + (m\omega'/2p_z^2)^{\frac{1}{2}}\Theta_n^{\pm'}\right]^{-1}\Big\} + (\Omega \to -\Omega)$$

where

$$-a_{rn}(\pm\pm') = \left[\pm 1 \pm' n - (n+2r+1)\right]\hbar\omega_c/2$$

and

$$\Theta_n^{\pm'} = \Omega - \pm' n\omega_c - \hbar p_z^2/2m \ .$$

This is equivalent to the result obtained by M.J. Stephen [1]. It should be noted that in the degenerate case the ω'-integral is of the form, (η_+ is a cutoff function),

$$\int_0^{\infty} d\omega' f_0(\omega' + a_{rn}(\pm\pm'))\ldots \to \eta_+(\zeta - a_{rn}(\pm\pm'))\int_0^{(\zeta - a_{rn}(\pm\pm'))} d\omega' \ldots \ .$$

Thus, in the degenerate case the individual terms of the series have the role of the chemical potential played by $(\zeta - a_{rn}(\pm\pm'))$ instead of (ζ), and are accompanied by a cutoff factor $\eta_+(\zeta - a_{rn}(\pm\pm'))$ which cuts off the series in the strong field case. In fact, in the quantum strong field limit, one has $\hbar\omega_c > \zeta$ as well as

$\hbar\omega_c\beta \gg 1$, and only one term of $\sum_{r=0}^{\infty} \sum_{\pm,\pm'}$, survives the cutoff prescription. (This convenience is the reason for introducing the series expansion $\sum_r$ above). If the temperature is not exactly zero in the degenerate case, the terms which are "cut-off" do not vanish entirely, but are small being proportional to the small factor $\exp\left[(\zeta - a_{rn}(\pm\,\pm'))\beta\right]$ because the ω'-integral is of the form $\int_0^{\infty} d\omega' f_0(\omega' + a_{rn}(\pm\,\pm'))\ldots$.

One can amplify on the remarks above by performing the ω'-integration in the degenerate case. The integral is elementary [6] and the result is given by

$$1 = \frac{4\pi e^2}{p^2}\frac{m^2\omega_c}{4\pi^2 p_z}\sum_{n=0}^{\infty}{}' \sum_{r=0}^{\infty}\left(\frac{\hbar\bar{p}^2}{2m\omega_c}\right)^n \exp\left(\frac{-\hbar\bar{p}^2}{2m\omega_c}\right)\frac{r!}{(n+r)!}\left[L_r^n\left(\frac{\hbar\bar{p}^2}{2m\omega_c}\right)\right]^2$$

$$\sum_{\pm,\pm'}\frac{1}{\hbar^3}\eta_+(\zeta - a_{rn}(\pm\pm'))\log\left|\frac{\Theta_n^{\pm'} + (2p_z^2/m)^{\frac{1}{2}}\left[\zeta - a_{rn}(\pm\pm')\right]^{\frac{1}{2}}}{\Theta_n^{\pm} - (2p_z^2/m)^{\frac{1}{2}}\left[\zeta - a_{rn}(\pm\pm')\right]^{\frac{1}{2}}}\right|$$

$$+ (\Omega \to -\Omega). \qquad (2.5)$$

In the quantum strong field limit ($\hbar\omega_c > \zeta$), the cutoff function may be taken as $\eta_+(\zeta - a_{rn}(\pm\pm')) \sim \delta(\pm'=+)\delta(\pm=+)\delta(r=0)$ (where the δ-functions are Kroenecker deltas). Noting that $L_0^n(x) \equiv 1$, one finds,

$$1 = \frac{m\omega_p^2}{2\hbar p^2}\left(\frac{m}{2p_z^2\zeta}\right)^{\frac{1}{2}} \exp\left(\frac{-\hbar\bar{p}^2}{2m\omega_c}\right)\sum_{n=0}^{\infty}\frac{1}{n!}\left(\frac{\hbar\bar{p}^2}{2m\omega_c}\right)^n .$$

$$\cdot\left\{\log\left|\frac{\Omega - n\omega_c - \hbar p_z^2/2m + (2p_z^2\zeta/m)^{\frac{1}{2}}}{\Omega - n\omega_c - \hbar p_z^2/2m - (2p_z^2\zeta/m)^{\frac{1}{2}}}\right| + (\Omega \to -\Omega). \qquad (2.6)\right.$$

Here $\omega_p^2 = 4\pi e^2 \rho/m$, and the appropriate expression for the density ρ in the quantum strong field limit is given by

$$\rho = m^{3/2}\, \omega_c \zeta^{1/2}/2^{1/2}\pi^2\, \hbar^2 \,. \tag{2.7}$$

This is a specialization of the more general expression for ρ in the degenerate case, which is given by

$$\rho = (m^{3/2}\, \omega_c/2^{1/2}\, \pi^2\, \hbar^2)$$

$$\sum_{\pm} \sum_{r=0}^{\infty} \left[\zeta \pm \tfrac{1}{2}\hbar\omega_c - (r+\tfrac{1}{2})\hbar\omega_c\right]^{\frac{1}{2}} \eta_+\left[\zeta \pm \tfrac{1}{2}\hbar\omega_c - (r+\tfrac{1}{2})\hbar\omega_c\right]. \tag{2.7$'$}$$

The relative importance of a root $\Omega(\vec{p})$ of the dispersion relation (2.1) (or (2.6) in the quantum strong field limit) in response to excitation is measured by the amplitude weight function $Z(\Omega(\vec{p}))$ which is given in reference (4d) by

$$Z(\Omega(\vec{p}))^{-1} = -\left[\frac{d}{d\Omega}\left(\begin{matrix}\text{right hand side of the}\\ \text{dispersion relation}\end{matrix}\right)\right]_{\Omega=\Omega(\vec{p})} \tag{2.8}$$

The damping $\gamma(\Omega(\vec{p}))$ of a mode or resonance which arises as a root of the dispersion relation may be expressed as $\gamma = Z\Gamma$, where Z is the amplitude weight function given by (2.8) and Γ is given by (reference 4d, eq. IV.1)

$$\Gamma(\vec{p},\Omega) = \frac{4\pi e^2}{p^2}\left(\frac{m}{2\pi}\right)^{3/2} \omega_c \int_0^{\infty} d\omega\, \frac{f_0(\omega)}{\hbar^3} \int_{c-i\infty}^{c+i\infty} \frac{ds}{2\pi i}\, e^{s\omega} \int_{-\infty}^{\infty} dy e^{-i\Omega y/2}$$

$$\cdot\ s^{-\frac{1}{2}} \frac{\sinh\,(\hbar\Omega s/2)}{\tanh\,(\hbar\omega_c s/2)}\ \exp\left[\frac{-p_z^2}{8ms}\,(y^2+\hbar^2 s^2)\right]$$

$$\exp\left[\frac{-\hbar\bar{p}^2}{2m\omega_c}\ \frac{\cosh\,(\hbar\omega_c s/2) - \cos(\omega_c y/2)}{\sinh\,(\hbar\omega_c s/2}\right]. \qquad (2.9)$$

Once again one may introduce an expansion in terms of modified Bessel functions $I_n(\sum_n)$, and then expand the latter in a series of the squares of Laguerre polynomials $L_r^n(\sum_r)$. This results in a y-integral which is simply evaluated as the Fourier transform of a Gaussian; and from this it is readily found that the s-integration gives rise to $\delta\,(\omega - a_{rn}(\pm\,\pm') - m\,(\Theta_n^{\pm'})^2/2p_z^2)$-functions so that the ensuing ω -integration is trivial. One finally obtains the result,

$$\Gamma(\vec{p},\Omega) = (8\pi e^2/p^2)\ (m/2\pi)^{3/2}\omega_c\ (8\pi m/p_z^2)^{\frac{1}{2}}$$

$$\sum_{n=0}^{\infty}{}' \sum_{r=0}^{\infty} \sum_{\pm} \sum_{\pm'} \frac{r!}{(n+r)!} \left(\frac{\hbar\bar{p}^2}{2m\omega_c}\right)^n$$

$$\exp\left(\frac{-\hbar\bar{p}^2}{2m\omega_c}\right)\left[L_r^n\left(\frac{\hbar\bar{p}^2}{2m\omega_c}\right)\right]^2$$

$$\left\{\frac{1}{4\hbar^3}\, f_0\left[a_{rn}(\pm\,\pm') + m(\Theta_n^{\pm'})^2/2p_z^2\right] - (\Omega\rightarrow -\Omega)\right. .$$

$$(2.10)$$

In the degenerate case, $f_0\left[a_{rn}(++') + \ldots\right] \rightarrow$ $\rightarrow \eta_+\left[\zeta - a_{rn}\,(\pm\,\pm') - \ldots\right]$, and the same sort of cutoff prescription that follows in the strong field case above occurs here also. Specifically, the quantum strong field limit for Γ is given by the formula

$$\Gamma(\vec{p},\Omega) = \frac{\pi m\omega_p^2}{\hbar p^2}\left(\frac{m}{2p_z^2\zeta}\right)^{\frac{1}{2}} \exp\left[\frac{-\hbar \bar{p}^2}{2m\omega_c}\right]\sum_{n=o}^{\infty}\frac{1}{n!}\left(\frac{\hbar \bar{p}^2}{2m\omega_c}\right)^n \cdot$$

$$\cdot\ \eta_+\left[\zeta - (\Omega - n\omega_c - \hbar p_z^2/2m)^2\, m/2p_z^2\right. = (\Omega \to -\Omega). \tag{2.11}$$

3. PLASMON MODES AND RESONANCES IN THE QUANTUM STRONG FIELD LIMIT

In view of the fact that the quantum strong field dispersion relation (2.6) is a prototype of the more general dispersion relations (2.5) and (2.4) in the degenerate case, it is worthwhile to remark on its analyticity properties, which determine the nature of the plasmon resonance spectrum in the quantum strong field limit and thereby reflect on the character of the spectrum more generally in the degenerate case. The dispersion relation (2.6) is just the real part of the full dispersion equation given by

$$1 = \frac{m\omega_p^2}{2\hbar p^2}\left(\frac{m}{2p_z^2\zeta}\right)^{\frac{1}{2}} \exp\left[\frac{-\hbar \bar{p}^2}{2m\omega_c}\right]\sum_{n=o}^{\infty}\frac{1}{n!}\left(\frac{\hbar \bar{p}^2}{2m\omega_c}\right)^n$$

$$\left\{\log \frac{\Omega - n\omega_c - \hbar p_z^2/2m + (2p_z^2\zeta/m)^{\frac{1}{2}}}{\Omega - n\omega_c - \hbar p_z^2/2m - (2p_z^2\zeta/m)^{\frac{1}{2}}} + (\Omega \to -\Omega)\right. \tag{3.1}$$

The log-functions in (3.1) have pairs of branch points at $\Omega^{\pm}_{B.P.n}$ given by

$$\Omega^{\pm}_{B.P.n} = n\omega_c + \hbar p_z^2/2m \pm (2p_z^2\zeta/m)^{\frac{1}{2}} \tag{3.2}$$

and one has also to choose branches in order to uniquely define the functions. The choice of branches determines the imaginary part of (3.1) and an appropriate choice yields the damping quantity $\frac{1}{2}\Gamma(\vec{p},\Omega)$ (2.11), which is non-zero only in each of the intervals $[\Omega^-_{B.P.n},\Omega^+_{B.P.n}]$ associated with a given pair of branch points, and vanishes elsewhere on the real Ω-axis. Whereas the appropriate choice of branches must be independently prescribed along with (3.1), this choice is uniquely dictated in the Green's function formulation of the plasma oscillation problem and is embodied in the construction of the spectral function associated with the inverse dielectric function (see ref. 4d, pp. 10-13, also pp. 18-22). Having the fully defined dispersion relation (2.6) and damping quantity [11] at our disposal, we may analyze plasmon resonances for all Ω, including the intervals $[\Omega^-_{B.P.n},\Omega^+_{B.P.n}]$ as well as the intervals $[\Omega^+_{B.P.n},\Omega^-_{B.P.(n+1)}]$.

The intervals $[\Omega^-_{B.P.n},\Omega^+_{B.P.n}]$ of the real Ω-axis are the appropriate branch cuts of the full dispersion equation (3.1) which is analytic for all other Ω. As they are branch cuts, the intervals $[\Omega^-_{B.P.n},\Omega^+_{B.P.n}]$ were excluded from consideration in the study of plasmon resonances by Mermin and Canel (ref. 2 pp 265 ff). The imaginary part of the full dispersion equation (3.1), being non-trivially defined in terms of the damping quantity $\frac{1}{2}\Gamma(\vec{p},\Omega)$(2.11) on the branch cuts, may be expected to result in a non-trivial damping of any resonances found in the neighborhoods of the branch cuts (of course such complex roots of (3.1) lie on the side of the branch cuts where the imaginary part of (3.1) varies continuously rather than the side of discontinuous variation). In fact, the dispersion relation does indeed have roots in the neighborhood of each branch cut $[\Omega^-_{B.P.n},\Omega^+_{B.P.n}]$ and the conclusion of Mermin and Canel that there is just on resonance near each value of $n\omega_c$ should be modified to recognize the presence of two resonances near $n\omega_c$, one of which lies in the neighborhood of the branch cut. We shall investigate the nature of these resonances in detail in order to ascertain their significance.

Our considerations are directed at the case in which the branch cuts do not overlap; in the quantum strong field limit this means that

$$\left[\text{Branch Cut Width}\right] = (8p_z^2\zeta/m)^{\frac{1}{2}} < \omega_c. \qquad (3.3)$$

Moreover, the presence of the logarithmic singularities in the quantum strong field dispersion relation (2.6) means that the curve representing the righthand side of (2.6) approaches infinity positively as Ω approaches a branch point $\Omega^+_{B.P.n}$ and it approaches infinity negatively as Ω approaches a branch point $\Omega^-_{B.P.n}$. (Obviously, similar remarks are valid for the dispersion relations (2.5) and (2.4) in the degenerate regime so long as the sharply "cut-off" character of the Fermi function is felt.) Since the curve representing the right-hand side of (2.6) is continuous between branch points, we are assured of at least one resonance in each of the intervals $\left[\Omega^+_{B.P.n}, \Omega^-_{B.P.(n+1)}\right]$, which must lie close to one of the branch points when wavenumber is small in the sense that $\hbar p^2/2m\omega_c \ll 1$.* Along the branch cuts $\left[\Omega^-_{B.P.n}, \Omega^+_{B.P.n}\right]$, there is not complete assurance that the increasing character of the curve is monotonic (and, therefore, it is conceivable that there may be more than one root along some of the branch cuts); however, under a wide variety of circumstances the curve does indeed

---

* This has already been pointed out by Mermin and Canel (ref. 2 pg. 265 ff). There seem to be some typographical errors in this reference.

increase monotonically along the branch cuts, and there is then just one root along each branch cut. For example, this is the case whenever the behavior of the curve along a branch cut is determined principally by the singularities at the branch points at the ends of that branch cut, with negligible interference from the singularities associated with the other branch cuts; again, the root lies close to one of the branch points when $\hbar\bar{p}^2/2m\omega_c \ll 1$.

Additional structure occurs in the curve representing the right-hand side of the dispersion relation along the branch cuts as the magnetic field is decreased below the value appropriate to the quantum strong field limit; that is when $\hbar\omega_c < \zeta$. (We consider only mild changes in this regard, say $\hbar\omega_c \lesssim \zeta < 2\hbar\omega_c$; of course, the curve would be radically altered if $\hbar\omega_c \ll \zeta$ but we do not consider this here). The additional structure arises from the presence of additional logarithmic singularities which give rise to additional branch cuts of width $\left[8p_z^2(\zeta-\hbar\omega_c)/m\right]^{\frac{1}{2}}$, which are superposed upon the branch cuts discussed above (having the same centers). From the degenerate dispersion relation (2.5) we see that these enter when $\hbar\omega_c \lesssim \zeta \leq 2\hbar\omega_c$ with relative strength $\left[L^n_{r=1}\ (\hbar\bar{p}^2/2m\omega_c)\right]^2$, and they also contribute to the damping quantity $\Gamma(\vec{p},\Omega)$ (2.10) but only along the additional roots (resonances), one outside each additional superposed branch cut, and one inside as well (which is rather more heavily damped). In the case of low wavenumber ($\hbar\bar{p}^2/2m\omega_c \ll 1$, so that all roots lie very close to their corresponding branch points), the excitation amplitudes of these additional roots are lower than the amplitudes of the roots discussed above (the latter roots being near $\Omega^{\pm}_{B.P.n}$) by a factor of $(1-\hbar\omega_c/\zeta)^{\frac{1}{2}}$.

Moreover, the influence that the additional singularities can have on the roots near $\Omega^{\pm}_{B.P.n}$ must be small in the sense that $(1-\hbar\omega_c/\zeta)^{\frac{1}{2}}$ is small. As we are concerned here with the quantum strong field limit $\hbar\omega_c > \zeta$, such effects are eliminated by the cutoff function $\eta_+(\zeta-\hbar\omega_c)$ in (2.5), which means that they are small in the same sense that $\exp\left[(\zeta-\hbar\omega_c)\beta\right]$ is small. However, it is well to bear in mind that such effects can become important as the magnetic field is decreased. (Further additional structure of the type described here is introduced into the curve representing the right-hand side of the dispersion relation as the magnetic field is further decreased such that the inequality $n\hbar\omega_c \lesssim \zeta$ holds for successively larger integers n. For each value of n achieved in this way through a reduction of ω_c, further additional branch cuts of width $\left[8p_z^2\,(\zeta-n\hbar\omega_c)/m\right]^{\frac{1}{2}}$ are superposed on the branch cuts appropriate to all lower integers $k < n$ which are already present (same centers); and there is a consequent increase in the number of roots.

Our quantitative considerations will be restricted to the quantum strong field limit ($\hbar\omega_c > \zeta$), in which case the dispersion relation may be written as

$$1 = \sum_{n=0}^{\infty} C_n(\vec{p}) \log\left|\frac{\Omega^2 - [\Omega^-_{BPn}]^2}{\Omega^2 - [\Omega^+_{BPn}]^2}\right| \tag{3.4}$$

where we have set

$$C_n(\vec{p}) = \frac{m\omega_p^2}{2\hbar p^2}\left(\frac{m}{2p_z^2\zeta}\right)^{\frac{1}{2}} \exp\left(\frac{-\hbar\bar{p}^2}{2m\omega_c}\right)\frac{1}{n!}\left(\frac{\hbar\bar{p}^2}{2m\omega_c}\right)^n . \tag{3.5}$$

(All the corrections due to additional structure of the type described in the preceding paragraph are vanishingly small in the sense that $\exp\left[(\zeta-\hbar\omega_c)\beta\right] << 1$.) A description of the principal modes and their associated structure may be obtained in the case when wavenumber is low in all possible senses by retaining only the n=0 and n=1 terms of (2.6). In the limit of zero wavenumber the n=0 term correctly reproduces the parallel propagation part of the dispersion relation ($\omega_p^2 \sin^2\theta/\Omega^2$) and the n=1 term correctly reproduces the perpendicular propagation part ($\omega_p^2 \cos^2\theta/\Omega^2-\omega_c^2$). There are three special cases in which the roots of the resulting zero-wavenumber dispersion relation may be obtained by treating the n=0 and n=1 terms independently; these cases are (i) $\theta = 0$, (ii) $\theta = \pi/2$, (iii) $\omega_p << \omega_c$ (all θ). When none of these three conditions is satisfied, the dispersion relation must be treated more generally, in a way that takes account of the n=0 and n=1 terms jointly; the two principal plasma modes which are exact solutions of the full zero-wavenumber dispersion relation,

$$1 = \omega_p \sin^2\theta/\Omega_0^2 + \omega_p^2 \cos^2\theta/\Omega_0^2 - \omega_c^2 \qquad (3.6)$$

(θ = angle between $\vec{p}$ and the plane perpendicular to $\vec{H}$) are described in reference 4d, eq. III.13ff. The result is

$$\Omega_{0\pm}^2 = \tfrac{1}{2}(\omega_p^2+\omega_c^2) \pm \tfrac{1}{2}\left[(\omega_p^2+\omega_c^2)^2 - 4\omega_p^2\omega_c^2\sin^2\theta\right]^{\frac{1}{2}} \qquad (3.7)$$

and a full account of the excitation amplitudes and damping of these principal modes is given in this reference also.

The zero wavenumber dispersion relation (3.6) describes the situation in which each of the wo pairs of branch points at $\Omega = \hbar p_z^2/2m \pm (2p_z^2\zeta/m)^{\frac{1}{2}}$ and $\Omega = \omega_c + \hbar p_z^2/2m \pm (2p_z^2\zeta/m)^{\frac{1}{2}}$ coalesces to give simple poles at $\Omega=0$ and $\Omega=\omega_c$ respectively. When wavenumber is small but nontrivial, the finite widths of the branch cuts associated with each pair of branch points must be taken into account. In accordance with (3.3) the widths are taken to be small so that there is essentially no interference between different branch cuts in the low wavenumber analysis (apart from that which occurs in the principal modes as described above). The structure of the plasma oscillation spectrum which occurs when the finite widths of the branch cuts are taken into account has already been mentioned; for each branch cut there are two plasmon resonances *, one of which lies alongside the branch cuts $[\Omega^-_{B.P.0},\Omega^+_{B.P.0}]$ and $[\Omega^-_{B.P.1},\Omega^+_{B.P.1}]$ respectively.

In order to calculate the resonances Ω'_{0+} and Ω'_{0-} associated with the principal modes, we retain only the n=0 and n=1 terms in (2.6) with the result

$$1 = C_0(\vec{p})\log\left|\frac{\Omega^2-[\hbar p_z^2/2m-(2p_z^2\zeta/m)^{\frac{1}{2}}]^2}{\Omega^2-[\hbar p_z^2/2m+(2p_z^2\zeta/m)^{\frac{1}{2}}]^2}\right|$$

$$+C_1(\vec{p})\log\left|\frac{\Omega^2-[\omega_c+\hbar p_z^2/2m-(2p_z^2\zeta/m)^{\frac{1}{2}}]^2}{\Omega^2-[\omega_c+\hbar p_z^2/2m+(2p_z^2\zeta/m)^{\frac{1}{2}}]^2}\right| \qquad (3.8)$$

---

*A convenient notation for the resonance which lies along the branch cut $[\Omega^-_{B.P.n},\Omega^+_{B.P.n}]$ is $\Omega'_{(n\omega_c)}$, and we denote the nearest resonance which lies outside the branch cut by $\Omega_{(n\omega_c)}$. The principal modes Ω_{0+} and Ω_{0-} are exceptions to this notation and the two resonances associated with the principal modes (resonances which are located along the branch cuts $[\Omega^-_{B.P.0},\Omega^+_{B.P.0}]$ and $[\Omega^-_{B.P.1},\Omega^+_{B.P.1}]$ will be denoted by Ω'_{0+} and Ω'_{0-}.

The principal modes themselves (Ω_{0+} and Ω_{0-}) may be understood as the two roots of (3.8) which lie outside the branch cuts $[\Omega^-_{B.P.o}, \Omega^+_{B.P.o}]$ and $[\Omega^-_{B.P.1}, \Omega^+_{B.P.1}]$ and, except in the special cases mentioned above, they must be calculated by taking account of the n=0 and n=1 terms jointly, with the zero-wavenumber limit given by (3.6,7). Whereas the n=0 and n=1 terms must be treated jointly in calculating the principal modes $\Omega_{0\pm}$, it is quite reasonable to assume that either associated resonance ($\Omega'_{0+}, \Omega'_{0-}$) is determined essentially by the branch cut which it lies beside. Accordingly, we shall solve (3.8) for $\Omega'_{0\{\pm\}}$ by treating the $n = \left\{ {0 \atop 1} \right\}$ term exactly, and giving only cursory treatment to the $n=\left\{ {1 \atop 0} \right\}$ term by replacing it by its zero-wavenumber limit evaluated at $\Omega = \left\{ {0 \atop 1} \right\} \omega_c$. Thus we write the dispersion relation (3.8) for Ω'_{0-} as follows,

$$1=C_0(\vec{p}) \log \frac{\Omega'^2_{0-} - [\hbar p_z^2/2m - (2p_z^2\zeta/m)^{\frac{1}{2}}]^2}{-\Omega'^2_{0-} + [\hbar p_z^2/2m + (2p_z^2\zeta/m)^{\frac{1}{2}}]^2} - \frac{\omega_p^2 \cos^2\theta}{\omega_c^2}, \tag{3.9}$$

and the solution is given by

$$\Omega'^2_{0-} = (\hbar p_z^2/2m)^2 + (2p_z^2\zeta/m)$$

$$+ (\hbar p_z^2/m)(2p_z^2\zeta/m)^{\frac{1}{2}} \tanh\left[(2C_0(\vec{p}))^{-1}(1+\omega_p^2\cos^2\theta/\omega_c^2)\right] \tag{3.10}$$

Similar considerations apply to Ω'_{0+} which is determined by

$$1=\frac{\omega_p^2\sin^2\theta}{\omega_c^2} + C_1(\vec{p}) \log \frac{\Omega'^2_{0+} - [\omega_c + \hbar p_z^2/2m - (2p_z^2\zeta/m)^{\frac{1}{2}}]^2}{-\Omega'^2_{0+} + [\omega_c + \hbar p_z^2/2m + (2p_z^2\zeta/m)^{\frac{1}{2}}]^2} \tag{3.11}$$

and the solution is given by

$$\Omega'^{2}_{0+} = (\omega_c + \hbar p_z^2/2m)^2 + (2p_z^2\zeta/m)$$

$$+ 2(2p_z^2\zeta/m)^{\frac{1}{2}} (\omega_c+hp_z^2/2m) \tanh\left[(2C_1(\vec{p}))^{-1}(1-\omega_p^2\sin^2\theta/\omega_c^2)\right] \tag{3.12}$$

The amplitude weight functions $Z(\Omega'_{0\pm})$ may be calculated with the use of equation (2.8) and introducing the appropriate approximations to the dispersion relation as given by (3.9) and (3.11) one obtains the results,

$$Z(\Omega'_{0-}) = -(\hbar^2p^2p_z^4\zeta/m^3\omega_p^2)\exp(h\bar{p}^2/2m\omega_c)$$

$$\cdot\left\{\Omega'_{0-} \cosh^2\left[(2C_0(\vec{p}))^{-1}(1+\omega_p^2 \cos^2\theta/\omega_c{}^2)\right]\right\}^{-1} \tag{3.13}$$

and

$$Z(\Omega'_{0+}) = -(4p^2p_z^2\zeta\,\omega_c/\bar{p}^2m\omega_p^2)(\omega_c+\hbar p_z^2/2m)\exp(h\bar{p}^2/2\,m\omega_c)$$

$$\cdot\left\{\Omega'_{0+} \cosh^2\left[(2C_1(\vec{p}))^{-1}(1-\omega_p^2\sin^2\theta/\omega_c^2)\right]\right\}^{-1}. \tag{3.14}$$

The impossibility of a resonance occuring exactly at one of the branch points $\Omega^{\pm}_{B.P.n}=n\omega_c+\hbar p_z^2/2m \pm(2p_z^2\zeta/m)^{\frac{1}{2}}$ is manifested in the cases treated here (n=0,1) by the fact that the amplitude weight functions would vanish if a resonance were to lie exactly on a branch point. (In the cases treated here, a resonance located at a branch point would correspond to $\tanh[\ldots]\rightarrow\pm 1$, and then $\cosh[\ldots]\rightarrow\infty$ so that $Z\rightarrow 0$.) In general, the branch points are forbidden as resonance frequencies.

A determination of the damping of Ω'_{0-} and Ω'_{0+} may be had by employing equation (2.11) with the results

$$\Gamma(\Omega'_{0-}) = (\pi m\omega_p^2/\hbar p^2)\ (m/2p_z^2\zeta)^{\frac{1}{2}}\ \exp\ (-\hbar\bar{p}^2/2m\omega_c) \quad (3.15)$$

and

$$\Gamma(\Omega'_{0+}) = (\pi m\omega_p^2/\hbar p^2)(m/2p_z^2\zeta)^{\frac{1}{2}}(h\bar{p}^2/2m\omega_c)\exp(-h\bar{p}^2/2m\omega_c). \quad (3.16)$$

Introducing these results into the damping formula $\gamma = Z\Gamma$, we have

$$\gamma(\Omega'_{0-}) = -\ (\pi\hbar p_z^2/2m)\ (2p_z^2\zeta/m)^{\frac{1}{2}}$$

$$\cdot\left\{\Omega'_{0-}\ \cosh^2\left[(2C_0(\vec{p}))^{-1}(1+\omega_p^2\ \cos^2\theta/\omega_c{}^2)\right.\right.^{-1} \quad (3.17)$$

and

$$\gamma(\Omega'_{0+}) = -\pi(\omega_c+\hbar p_z^2/2m)\ (2p_z^2\zeta/m)^{\frac{1}{2}}$$

$$\cdot\left\{\Omega'_{0+}\ \cosh^2\left[(2C_1(\vec{p}))^{-1}(1-\omega_p^2\ \sin^2\theta/\omega_c{}^2)\right]\right\}^{-1} \quad (3.18)$$

The calculation of the resonance $\Omega'_{(n\omega_c)}$ which lies along the branch cut $[\Omega^-_{B.P.n}, \Omega^+_{B.P.n}]$, and of the resonance $\Omega_{(n\omega_c)}$ which lies just outside this branch cut may be carried out in a similar manner. Assuming that $\Omega'_{(n\omega_c)}$ are determined essentially by the singularities which this branch cut is associated with, one retains only the nth term in (2.6) and approximates the n=0 and n=1 terms by their zero-wavenumber limit evaluated at $\Omega = n\omega_c$. Thus the resonances $\Omega'_{(n\omega_c)}$ and $\Omega_{(n\omega_c)}$ are determined by the approximation (3.19) to the dispersion relation,

$$1=\frac{\omega_p^2\sin^2\theta}{(n\omega_c)^2}+\frac{\omega_p^2\cos^2\theta}{(n\omega_c)^2-\omega_c^2}$$

$$+\ C_n(\vec{p})\ \log\left|\frac{\Omega^2-[n\omega_c+\hbar p_z^2/2m-(2p_z^2\zeta/m)^{\frac{1}{2}}]^2}{\Omega^2-[n\omega_c+\hbar p_z^2/2m+(2p_z^2\zeta/m)^{\frac{1}{2}}]^2}\right| \tag{3.19}$$

Recognizing that for $\Omega\to\Omega_{(n\omega_c)}$ one may replace the absolute value in the log-term by its argument, whereas for $\Omega\to\Omega'_{(n\omega_c)}$ the absolute value must be replaced by the negative of its argument, one may solve (3.19) for the resonances $\Omega'_{(n\omega_c)}$ and $\Omega_{(n\omega_c)}$ with the following results,

$$\Omega'^2_{(n\omega_c)}=(n\omega_c+\hbar p_z^2/2m)^2+(2p_z^2\zeta/m)+(8p_z^2\zeta/m)^{\frac{1}{2}}(n\omega_c+\hbar p_z^2/2m)$$

$$\tanh\left[(2C_n(\vec{p}))^{-1}(1-\omega_p^2\sin^2\theta/(n\omega_c)^2-\omega_p^2\cos^2\theta/(n\omega_c)^2-\omega_c^2)\right] \tag{3.20}$$

and

$$\Omega^2_{(n\omega_c)}=(n\omega_c+\hbar p_z^2/2m)^2+(2p_z^2\zeta/m)+(8p_z^2\zeta/m)^{\frac{1}{2}}(n\omega_c+\hbar p_z^2/2m)$$

$$\cdot\ \mathrm{ctnh}\left[(2C_n(\vec{p}))^{-1}(1-\omega_p^2\sin^2\theta/(n\omega_c)^2-\omega_p^2\cos^2\theta/(n\omega_c)^2-\omega_c^2)\right]. \tag{3.21}$$

The amplitude weight functions may be calculated with the use of equations (2.8) and (3.19) with the results,

$$Z(\Omega'_{(n\omega_c)})$$

$$= -\frac{(2\hbar p^2 p_z^2 \zeta/m^2\omega_p^2)(n\omega_c+\hbar p_z^2/2m)n!(2m\omega_c/\hbar\bar{p}^2)^n \exp(\hbar\bar{p}^2/2m\omega_c)}{\Omega'_{(n\omega_c)}\cosh^2\left[(2C_n(\vec{p}))^{-1}(1-\omega_p^2\sin^2\theta/(n\omega_c)^2-\omega_p^2\cos^2\theta/(n\omega_c)^2-\omega_c^2)\right]}$$

(3.22)

$$Z(\Omega_{(n\omega_c)})$$

$$= \frac{(2\hbar p^2 p_z^2 \zeta/m^2\omega_p^2)(n\omega_c+\hbar p_z^2/2m)n!(2m\omega_c/\hbar\bar{p}^2)^n \exp(\hbar\bar{p}^2/2m\omega_c)}{\Omega_{(n\omega_c)}\sinh^2\left[(2C_n(\vec{p}))^{-1}(1-\omega_p^2\sin^2\theta/(n\omega_c)^2-\omega_p^2\cos^2\theta/(n\omega_c)^2-\omega_c^2)\right]}$$

(3.23)

Since the damping quantity $\Gamma(\vec{p},\Omega)$ is nonzero only along the branch cuts, the resonance $\Omega_{(n\omega_c)}$ is undamped (to the extent that its damping is exponentially small in the sense that $\exp\left[(\zeta-h\omega_c)\beta\right] \ll 1$. On the other hand, the resonance $\Omega'_{(n\omega_c)}$ along the branch cut is damped, and the value of $\Gamma(\Omega'_{(n\omega_c)})$ is given by

$$\Gamma(\Omega'_{(n\omega_c)})$$

$$= (\pi m\omega_p^2/\hbar p^2)(m/2p_z^2\zeta)^{\frac{1}{2}}(1/n!)(\hbar\bar{p}^2/2m\omega_c)^n \exp(-\hbar\bar{p}^2/2m\omega_c).$$

(3.24)

The damping constant $\gamma(\Omega'_{(n\omega_c)})=Z(\Omega'_{(n\omega_c)})\Gamma(\Omega'_{(n\omega_c)})$ is then given by

$$\gamma(\Omega'_{(n\omega_c)})$$

$$= -\frac{\pi(n\omega_c + \hbar p_z^2/2m)(2p_z^2\zeta/m)^{\frac{1}{2}}}{\Omega'_{(n\omega_c)}\cosh^2\left[(2C_n(\vec{p}))^{-1}(1-\omega_p^2\sin^2\theta/(n\omega_c)^2-\omega_p^2\cos^2\theta/(n\omega_c)^2-\omega_c^2)\right]}$$

(3.25)

It should be noted that the daming of the resonance $\Omega'_{\frac{1}{2}(n\omega_c)}$ (3.25) is small in the same sense that $(2p_z^2\zeta/m)^{\frac{1}{2}}$ is small (in units of frequency), so that the resonance $\Omega'_{(n\omega_c)}$ may play a role in physical response, although the undamped resonance $\Omega_{(n\omega_c)}$ will be more important. Finally, we note that the calculated damping constant $\gamma(\Omega'_{(n\omega_c)})$ (3.25) is negative. This should be no surprise since the amplitude weight function $Z(\Omega'_{(n\omega_c)})$ (which is involved in calculating $\gamma = Z\Gamma$) is also negative, whereas $Z(\Omega_{(n\omega_c)})$ is positive. There is no difficulty in connection with this fact because the associated Lorentzian contribution to the spectral function is of the form

$$\bar{a}\,(\vec{p},\omega) \qquad \frac{Z(\Omega'_{(n\omega_c)})\,\gamma(\Omega'_{(n\omega_c)})}{(\omega-\Omega'_{(n\omega_c)})^2+(\frac{1}{2}\gamma(\Omega'_{(n\omega_c)}))^2}$$

and this is insensitive to reversals of the signatures of $Z(\Omega'_{(n\omega_c)})$ and $\gamma(\Omega'_{(n\omega_c)})$ when such sign reversals are taken together. Thus one should simply understand $Z(\Omega'_{(n\omega_c)})$ and $\gamma(\Omega'_{(n\omega_c)})$ to be replaced by their magnitudes.

The appropriate expression for the dispersion relation for perpendicular plasmon propagation in the quantum strong field limit may be written in an interesting closed form by means of (2.1) and (2.2). An expansion of the integrands of (2.1) and (2.2) in a series powers of $\exp(-\hbar\omega_c s/2)$ yields the desired result when all terms other than the leading one are neglected; thus, in the integrands of (2.1) and (2.2) one may set

$$\frac{\exp\left[(-\hbar\bar{p}^2/2m\omega_c)\coth(\hbar\omega_c s/2)\right]}{\tanh(\hbar\omega_c s/2)}$$

$$\exp\left[\frac{\hbar\bar{p}^2}{2m\omega_c}\;\frac{\cos\left[(\omega_c/2)(2T-i\hbar s)\right]}{\sinh(\hbar\omega_c s/2)}\right] \rightarrow \exp\left[\frac{\hbar\bar{p}^2}{2m\omega_c}\;(e^{i\omega_c T}-1)\right].$$

With the appropriate identification of ρ and $\omega_p{}^2$, one obtains the result, (p_z=o),

$$1=\frac{m\omega_p^2}{\hbar p^2}\, i\int_0^\infty dT e^{-i\Omega T}\exp\left[\frac{\hbar\bar{p}^2}{2m\omega_c}\;(e^{i\omega_c T}-1)\right] + (\Omega\rightarrow-\Omega). \tag{3.26}$$

The integral $\int_0^\infty dT$ may be rewritten changing the integration variable to $\tau=\omega_c T$, and breaking up the integration range $[0,\infty]$ into intervals $[2\pi n,\ 2\pi(n+1)]$. Each such interval may be translated to the origin, with the occurrence of a factor $\exp(-i2\pi n\Omega/\omega_c)$ which goes into $[1-\exp(-i2\pi\Omega/\omega_c)]^{-1}$ when the summation over all such intervals is carried out. Further manipulation of this sort yields the result,

$$\int_0^\infty dT\ldots=\frac{\exp(-\hbar\bar{p}^2/2m\omega_c)}{i\omega_c\sin(\pi\Omega/\omega_c)}\int_0^\pi d\tau'\exp\left(-\frac{\hbar\bar{p}^2}{2m\omega_c}\cos\tau'\right)$$

$$\cos\left(\frac{\Omega}{\omega_c}\tau'+\frac{\hbar\bar{p}^2}{2m\omega_c}\sin\tau'\right),$$

and the integral on the right hand side may be identified in terms of the incomplete gamma function $\gamma\left(-\frac{\Omega}{\omega_c};\,-\frac{\hbar\bar{p}^2}{2m\omega_c}\right)$ (B.H.T.F.[5] II, pg. 137 2), with the result

$$\int_0^\infty dT \ldots = \frac{i}{\omega_c}\left[\exp\left(\frac{-\hbar\bar{p}^2}{2m\omega_c}\right)\right]\left(\frac{-\hbar\bar{p}^2}{2m\omega_c}\right)^{\Omega/\omega_c}\gamma\left(\frac{-\Omega}{\omega_c};\frac{-\hbar\bar{p}^2}{2m\omega_c}\right).$$

Thus we obtain a relatively simple closed expression for the case of perpendicular propagation, which can be rewritten in terms of the function $\gamma^*(-a;-x) = \frac{(-x)^a}{\Gamma(-a)}\gamma(-a;-x)$, which is a single-valued entire function of both a and x and is real for real a and x (B.H.T.F.[5] II, pg. 133, # 5), with the result,

$$1 = -(m\omega_p^2/\hbar\omega_c p^2)$$

$$\exp(-\hbar p^{-2}/2m\omega_c)\,\Gamma(-\Omega/\omega_c)\,\gamma^*(-\Omega/\omega_c;-\hbar\bar{p}^2/2m\omega_c) + (\Omega \to -\Omega) \tag{3.27}$$

There are simple poles at all integral values of (Ω/ω_c) as indicated by the presence of the gamma function, $\Gamma(-a) = -\pi\left[\sin\pi a\,\Gamma(1+a)\right]^{-1}$. Clearly this indicates that there is just one plasmon resonance near each value $n\omega_c$ in the special case of plasmon propagation perpendicular to the magnetic field, and equation (3.27) may be used to analyze the behavior of these resonances in the quantum strong field limit in both the cases $\hbar\bar{p}^2/2m\omega_c \ll 1$ and also $\hbar\bar{p}^2/2m\omega_c \gg 1$.

4. CONCLUSIONS

The examination of plasmon resonances which has been carried out here shows that there are two plasmon resonances near each multiple of the cyclotron frequency in the quantum strong field limit, $\Omega_{(n\omega_c)}$ (equation (3.21))and $\Omega'_{(n\omega_c)}$ (equation (3.20)). One of these $(\Omega_{(n\omega_c)})$ is undamped whereas the other $(\Omega'_{(n\omega_c)})$ is damped, and it has already been pointed out that this damping is small in the same sense that $(2p_z^2\zeta/m)^{\frac{1}{2}}$ is small. It is of interest to inspect the question of the relative importance of the two resonances $\Omega_{(n\omega_c)}$ and $\Omega'_{(n\omega_c)}$ further. The damping of $\Omega'_{(n\omega_c)}$ may be sensibly approximated by (see equation (3.25)),

$$|\gamma(\Omega'_{(n\omega_c)})| \cong \pi(2p_z^2\zeta/m)^{\frac{1}{2}}$$

$$\cosh^{-2}\left[(2C_n(\vec{p}))^{-1}(1-\omega_p^2\sin^2\theta/(n\omega_c)^2 - \omega_p^2\cos^2\theta/(n\omega_c)^2-\omega_c^2)\right]$$

where

$$(2C_n(\vec{p}))^{-1}$$

$$=(\hbar p^2/m\omega_p^2)(2p_z^2\zeta/m)^{\frac{1}{2}}\ n!\ (2m\omega_c/\hbar\bar{p}^2)^n \exp(\hbar\bar{p}^2/2m\omega_c).$$

That $\gamma(\Omega'_{(n\omega_c)})$ is small in the same sense that $(2p_z^2\zeta/m)^{\frac{1}{2}}$ is small is immediately obvious. In addition to this, the factor $\cosh^{-2}[\ldots] \sim e^{-2[\ldots]}$ causes $\gamma(\Omega'_{(n\omega_c)})$ to be exponentially small, under conditions of low wavenumber which imply that $(2C_n(\vec{p}))^{-1} \gg 1$ provided that propagation is not too close to the perpendicular direction. Thus the actual damping of $\Omega'_{(n\omega_c)}$ is very small indeed, albeit technically non-zero in comparison with the damping of $\Omega_{(n\omega_c)}$. A direct comparison of the relative excitation amplitudes of $\Omega'_{(n\omega_c)}$ and $\Omega_{(n\omega_c)}$ further supports the view that a long-wavelength analysis which takes into account of $\Omega_{(n\omega_c)}$ must also take account of $\Omega'_{(n\omega_c)}$. The ratio $|Z(\Omega'_{(n\omega_c)})/Z(\Omega_{(n\omega_c)})|$ may be calculated using (3.22) and (3.23), with the result,

$$\left| Z(\Omega'_{(n\omega_c)})/Z(\Omega_{(n\omega_c)}) \right| = (\Omega_{(n\omega_c)}/\Omega'_{(n\omega_c)})$$

$$\tanh^2 \left[(2C_n(\vec{p}))^{-1}(1-\omega_p^2\sin^2\theta/(n\omega_c)^2-\omega_p^2\cos^2\theta/(n\omega_c)^2-\omega_c^2) \right]$$

Under conditions of low wavenumber (as described above) we have $\tanh^2 \ldots \quad 1$ and then the excitation amplitudes of $\Omega'_{(n\omega_c)}$ and $\Omega_{(n\omega_c)}$ are seen to be comparable. One must conclude that the resonance $\Omega'_{(n\omega_c)}$ is comparable in importance to $\Omega_{(n\omega_c)}$ in terms of both excitation amplitude and damping under conditions of low wavenumber which imply that $(2C_n(\vec{p}))^{-1} \gg 1$ provided that propagation is not too close to the perpendicular direction; and accordingly $\Omega'_{(n\omega_c)}$ must be taken into account along with $\Omega_{(n\omega_c)}$ in physical response.

For plasmon propagation perpendicular to the magnetic field (or nearly perpendicular) there is just one resonance near each multiple of the cyclotron frequency in all regimes of magnetic field strength. In the quantum strong field limit, equation (3.27) may be used to analyze the behavior of these resonances in both the cases $\hbar\bar{p}^2/2m\omega_c \ll 1$ and also $\hbar\bar{p}^2/2m\omega_c \gg 1$.

REFERENCES

1) M.J. Stephen, Phys. Rev. 129, 997 (1963).
2) N.D. Mermin and E. Canel, Annals of Physics 26, 247 (1964).
3) A. Ron, Phys. Rev. 134, A70 (1964).
4) N.J. Horing, Ph.D. Thesis, Harvard University, (March 1964); Proc. Plasma Effects Symposium, Int. Conf. on Physics of Semiconductors, Paris (July 1964); Phys. Rev. 136, A494 (1964); Annals of Physics 31, 1 (1965).

5) a. B.H.T.F. refers to Bateman Manuscript Project, "Higher Trancendental Functions", Mc Graw Hill, 1953.

b. B.I.T. refers to Bateman Manuscript Project, "Tables of Integral Transforms", Mc Graw Hill, 1954.

6) Dwight, "Table of Integrals" [#188.11] , MacMillan Co., 1947.

Printed in the United States
By Bookmasters